Digital Security Field

Christopher Q

I0069033

Published by **Purple Team Security**

ISBN: 979-8-9988306-5-5

10 9 8 7 6 5 4 3 2

Digital Security Field Manual (DSFM)
Copyright © 2025 by Purple Team Security
All rights reserved.

ISBN: 979-8-9988306-5-5
Library of Congress Control Number: 2025911171

Second Edition
Published by Purple Team Security
Printed in the United States of America
Published on May 15, 2025

10 9 8 7 6 5 4 3 2

LEGAL DISCLAIMER

This manual, this **arsenal of defense**, this **blueprint for personal sovereignty**—is a work of education, not sedition. It does not deputize you as a digital Robin Hood, nor does it grant you license to dance gleefully on the wrong side of the law. If you've come here looking for a hall pass to commit crimes, let me assure you: **you are wrong.**

What you now hold—or perhaps, what now glares at you in digital form—is not a weapon. It is not a blueprint for criminal conquest. It is, rather, a **compendium of defensive knowledge**, a **tactical field guide** for those wishing to survive the predatory chaos of the digital world.

By continuing beyond this page, you accept that you are solely accountable for your actions, choices, and consequences. The author, publisher, and anyone remotely involved disclaim all liability for your inevitable rendezvous with law enforcement should you ignore this warning.

This book is for **defense**, not destruction. For **protection**, not exploitation. It is for those who walk the line between freedom and surveillance, privacy and exposure, right and ruin.

By proceeding, you acknowledge that you are of sound mind, or at the very least, that you can convincingly fake it in a court of law. You accept that this is a **toolbox**, not a ticket. A lesson, not a loophole. What you build with it is entirely, irrevocably, and—may the odds be ever in your favor—**your burden alone.**

Table of Contents

1.

The Story Behind This Manual

Every book starts somewhere. Some begin as research papers; others, as corporate whitepapers dressed up to look like something they're not. This one? It started as something far simpler: a checklist.

It was February 2025. I was preparing for a trip to Europe over spring break—nothing unusual, just a visit to Germany to watch my two favorite soccer clubs, Schalke 04 and 1. FC Nürnberg, take the pitch. But I wasn't about to leave my operational mindset at home. As an InfoSec professional, I knew bringing my laptop was the smart move. I understood the threat landscape here in the U.S., but I assumed Europe was similar. And as we all know, "assuming" isn't how you stay secure.

So, I started writing a list.

Privacy screen? Check.

Software firewall? Installed and configured.

Unnecessary services? Disabled.

LUKS full-disk encryption? Verified.

Nuke password for LUKS? Tested. Metaphorically,

just verified setup, for obvious reasons.
USB kill switch rigged to shut down my machine if
snatched in a café? Ready to go.

Somewhere between line item thirty and forty, I
paused and said aloud—half-joking, and half-not-
joking:

"Why isn't there a field manual for this?"

And that was it. The spark. The Digital Security
Field Manual was born—not as a book deal, not as
a thought leadership stunt—but as a tool I wished
already existed. Something real. Something usable.
Something you could actually apply the moment you
closed the book.

The first draft was done in five hours. It
wasn't pretty. It wasn't polished. But it worked.
What you're holding now is the refined second
edition—built from that original skeleton, expanded
with everything I've learned since, and shaped
by the realities faced not just by cybersecurity
professionals, but by everyday people, journalists,
activists, executives, and privacy-conscious
individuals worldwide. You should be able to
surf the web and not feel like a victim afterward.
You should be able to do your job without worrying
about being compromised. This book, while a great
start, will help you get there. However, this is
only the first step.

This isn't theory. It's practice. It's what I do,
and it's what I hope this book helps you do.

Stay safe. Stay sharp. Stay sovereign.

2.

About the Digital Security Field Manual

There was a time when privacy meant four walls and a locked door. Today, those walls are glass, the lock is an illusion, and someone's watching through the peephole with a high-powered telescope... or worse, a data broker's API.

Your personal data is under siege.

Governments spy in the name of national security. Corporations harvest your clicks, keystrokes, and even your silences—all in the name of "personalized experiences." Meanwhile, cybercriminals lurk in the shadows, sipping stale coffee while auctioning off your stolen data on forums you'll never visit.

Make no mistake: your life, your habits, your digital soul—they are *commodities*. Bought, sold, and weaponized. Not in some far-flung dystopian future, but here. Now. While you read these very words.

This manual—this humble little field guide—is your countermeasure. Not a silver bullet, but a loaded magazine.

Inside, you'll find tactics designed to:

- De-Google your smartphone, stripping it of corporate surveillance tentacles.
- Build air-gapped systems so isolated, even nation-states would blush.
- Deploy military-grade encryption because, frankly, "good enough" is not.
- Communicate like a ghost in the machine—present, but untraceable.
- Fortify your operating systems until even forensic analysts question their career choices.

You see, the internet is not a highway; it's a warzone. And in this war, information is the prize. Your information.

While the security world overflows with shiny tools and whispered techniques, this manual focuses on what works—tried, tested, and, most importantly, used by me. That said, don't treat these pages as gospel. Treat them as your *opening statement*. Research, adapt, evolve. The arms race never ends.

> **Your Privacy Matters**
>
> **Privacy is not a privilege—it's your birthright.** Every move you make to lock down your digital life is a blow against profiling, exploitation, and control.
>
> You are the guardian of your privacy—the first, the last, the only.

Threat actors, governments, corporations—they never sleep. Their tactics evolve daily. So, I ask you:

Will you be the product?

Or will you be the problem?

Let's begin.

3.

Introduction to Digital Security

Ah, the digital age. A marvel of convenience, a miracle of connection, and, dare I say, a masterclass in manipulation. Every click, every swipe, every whispered search in the dead of night—you're not alone. You're never alone.

Governments, corporations, cybercriminals—they all sit at the same table, feasting on data. *Your* data. Your habits, your movements, your weaknesses. Tracked, logged, sold, and served back to you with a smile.

It doesn't matter if you're a cautious citizen, a whistleblower on the run, or someone who just doesn't like the idea of a faceless algorithm knowing when you sleep, when you wake, and when you order pizza at 2 AM.

If you're reading this, you've already taken the first step. Welcome.

This guide is your roadmap to digital resistance. You'll learn how to:

- Break free from the clutches of corporate surveillance with a de-Googled smartphone.

- Build a computer so isolated it could make a submarine blush.

- Encrypt your data until even quantum computers throw their hands up.

- Vanish into the digital ether with anonymous communications.

- Fortify your systems against even the most curious forensic analyst.

Surveillance is not a conspiracy—it's an industry. And business, my friend, is booming.

Understanding Digital Security

Digital security, at its core, is an elegant cocktail of technology, behavior, and common sense. Three pillars hold it together:

- **Technical Security** - Encryption, hardened systems, and hardware that doesn't rat you out.

- **Operational Security (OPSEC)** - Your behavior. Your habits. Your discipline. The art of not being an easy mark.

- **Physical Security** - Because what good is a fortress if someone can just walk in through the front door?

Every action you take leaves a trace. This manual? It teaches you how to clean up after yourself.

Threat Modeling: Who Are You Protecting Against?

Let's not get ahead of ourselves. You can't defend what you haven't defined.

Take a moment. Ask yourself:

- Who's coming for you? Hackers? Corporations? Governments? A jealous ex?
- What's at stake? Your identity? Your finances? Your reputation? Your freedom?
- What's the most likely attack? Phishing? Malware? Physical surveillance? Social engineering?

Depending on your answers, your defense strategy changes.

- **Casual Privacy Seekers:** Block trackers, use encrypted apps, limit what you share.
- **Professionals & Activists:** Harden your devices, compartmentalize identities, expect targeted attacks.
- **High-Risk Operators:** Go offline. Air-gap. Operate like your life depends on it—because it might.

Common Digital Security Threats

Know your enemy. Here's the shortlist:

- **Mass Surveillance:** Metadata is the new gold. You are the mine.

- **Device Exploits:** Every unpatched vulnerability is an open door.

- **Social Engineering:** The easiest system to hack is the one between your ears.

- **Metadata Collection:** Encrypted or not, your patterns betray you.

- **Compromised Networks:** Free Wi-Fi? Free surveillance.

- **Supply Chain Attacks:** Sometimes, the trojan horse arrives shrink-wrapped and factory-sealed.

These threats aren't hypothetical—they're operational.

Best Practices for Digital Security

Let's cut the fluff. Here's what works:

1. Reduce Your Attack Surface

- Run minimalist operating systems (GrapheneOS, Qubes OS, Tails).

- Uninstall what you don't need. Bloatware is surveillance by another name.

- Avoid cloud services without end-to-end encryption. No exceptions.

2. Encrypt Everything

- Full-disk encryption on every device. No excuses.

- Use Signal, Briar, or Session for messaging. If it's not end-to-end encrypted, it's public.

- Lock down your sensitive files with VeraCrypt or GPG.

3. Strengthen Your OPSEC

- Never reuse passwords. Ever. Use Bitwarden or KeePassXC.

- Burn your email addresses like a spy burns an identity.

- Social media is a buffet for attackers. Serve them nothing.

4. Use Anonymity Tools

- Tor Browser. Mullvad Browser. *Become nobody.*

- Route your traffic through VPNs or Tor Bridges.

- Ditch biometric unlocks. Your face and fingerprints belong to you—not your phone.

What This Guide Covers

This manual is not a bedtime story. It's a blueprint for rebellion.

1. **Ultra-Secure Smartphones:** De-Google, harden, and control your communications.

2. **Air-Gapped Systems:** Build computers that know nothing of networks.

3. **Encryption Mastery:** Lock down your data, your emails, your life.

4. **Anonymous Internet Use:** Move through the digital world like a shadow.

5. **Operational Security:** Practice behaviors that leave attackers grasping at air.

By the end, you'll be armed with more than tools. You'll have a mindset.

Because privacy isn't just something you install. *It's something you become.*

4.

OSINT & Threat Intelligence

4.1 Understanding OSINT and Its Dangers

They say knowledge is power. Allow me to correct that.

Knowledge about you is power.

And every click, post, photo, or careless share—contributes to a dossier you never agreed to build.

Open-Source Intelligence, or OSINT if you're feeling punchy, is the art of profiling people using nothing but what they themselves—and their technologies—carelessly leave behind. No hacking. No backdoors. Just breadcrumbs. Millions of them.

It's used by security professionals. Law enforcement. Corporate spies. Political operatives. Hackers in dark basements with nothing better to do. And yes, that kid you blocked on social media last week.

How OSINT Is Used Against You:

• **Doxxing** - Your home, your phone number, your

email—packaged and posted for the world to see.

- **Social Engineering** - Attackers don't need to guess your mother's maiden name when you posted it on Facebook in 2011.

- **Targeted Cyber Attacks** - The more they know, the more precise their attacks become.

- **Corporate Espionage** - Think LinkedIn is boring? It's a goldmine for your competitors.

With nothing but public data, an adversary can map your life—your habits, your routines, your soft spots.

4.2 How Attackers Gather Information (OSINT Techniques)

OSINT isn't magic. It's methodical. It comes in two flavors:

- **Passive Reconnaissance** - The art of watching silently.

- **Active Reconnaissance** - The art of poking until something falls over.

1. Social Media Intelligence (SOCMINT):

- **Username Tracking** - Your clever handle? You've reused it everywhere.

- **Geotag Analysis** - Your photos tell them where you've been. Your timestamps tell them when.

- **Hashtag Monitoring** - You thought that meme was funny. They thought it connected you to twenty others just like you.

- **Metadata Extraction** - That picture you posted? It didn't just capture your smile. It captured your phone model, GPS coordinates, and the time you took it.

2. Domain & Infrastructure Reconnaissance:

- **WHOIS Lookups** - You registered that domain? They know who you are.

- **DNS Enumeration** - They map your digital empire, one subdomain at a time.

- **IP Geolocation** - They know where your server lives.

- **Port Scanning & Vulnerability Detection** - Ever heard of Shodan? It's Google for insecure devices. Spoiler: you're in it.

3. Data Breach & Dark Web OSINT:

- **Leaked Credentials** - That old password you used in 2015? Still floating around in the darker corners of the internet.

- **Dark Web Marketplaces** - You'd be amazed at what's for sale.

- **Tor Hidden Services** - The basement of the internet—full of intelligence, if you know where to look.

4. Search Engine Hacking (Google Dorking): The world's most powerful surveillance tool is free, fast, and starts with a G.

```
1   site:example.com filetype:pdf "confidential"
2   intitle:"index of" site:example.com
3   inurl:/wp-content/uploads/ site:example.com
4   site:pastebin.com intext:"password"
```

How to Block OSINT Reconnaissance:

- Use WHOIS privacy when registering domains.
- Hunt yourself down. Scan for leaked credentials. Change them—yesterday.
- Lock down your robots.txt to keep Google out of places it doesn't belong.
- Strip metadata from your photos before uploading. Every. Single. Time.

4.3 Defensive OSINT: Reducing Your Digital Exposure

You can't stop people from looking. But you can make sure they find nothing worth seeing.

Step 1: Remove Your Data from Public Databases

- Opt out of data brokers. Yes, it's a pain. Do it anyway.
- Use Google's removal tools to purge what you can.
- Burn down your old accounts with JustDeleteMe.

Step 2: Hide Your Online Identity

- Pseudonyms. Burner accounts. Your real name has no place in your public digital life.
- VPNs. Tor. Proxies. Your IP is not your identity—don't treat it like one.
- Never reuse usernames. Never post the same photo twice.

Step 3: Secure Your Digital Communications

- Signal. Session. Briar. Anything less is malpractice.

- Ditch Gmail. ProtonMail or self-hosted only.

- Encrypt your files like they're state secrets—because to you, they are.

Step 4: Break OSINT Correlation

- Live compartmentalized. Different names, different emails, different devices.

- Stop linking your social media like it's a business card.

- Change your habits. Regularly. Predictability is a gift you should never give.

4.4 Countering AI-Powered OSINT Tracking

Human OSINT is dangerous. AI OSINT is relentless.

AI Threats Include:

- **Facial Recognition** - Your face is a barcode. Clearview AI scans billions of them. Yours included.

- **Gait Analysis** - You have a digital "walk." AI knows it. Even if your face is hidden.

- **Metadata Correlation** - Machine learning stitches your digital life together—one post, one comment, one habit at a time.

How to Fight Back:

- Use Fawkes to poison facial recognition algorithms.

- Change your digital behaviors like you change your socks—often, and without notice.

- Stop posting unique photos. Yes, even that perfect latte art.

- Use burner accounts. Never tie your real identity to public engagement.

4.5 Final Thoughts: The Future of OSINT & Privacy Defense

The future belongs to those who can outmaneuver the machine. OSINT is getting smarter. So must you.

Build multiple digital identities. Encrypt everything. Make your trail look like static to the untrained eye—and to the algorithms.

This is not a one-time fix. *This is your new lifestyle.*

The most dangerous profiles are not the ones attackers can build. They're the ones they can't.

5.

Building an Ultra-Secure Smartphone

Your smartphone. The one buzzing in your pocket right now. A camera you can't see through. A microphone you don't control. A GPS tracker that never sleeps.

Congratulations. You're carrying the world's most efficient surveillance device. And you paid for it.

Modern smartphones are marvels of engineering—and absolute disasters for privacy. They know your location, your habits, your weaknesses. They track you when you sleep, when you travel, when you whisper secrets into a screen you trust far too much.

But you're not here to settle for that, are you?

This chapter will show you how to take back control—turning that pocket spy into a fortress of your own design.

5.1 Your Mission:

- Select hardware that works *for you*, not against you.

- Install a hardened, surveillance-free operating system.
- Lock down wireless vectors like Bluetooth and Wi-Fi.
- Harden your digital habits with ruthless precision.

5.2 Choosing the Right Phone

Not all phones are created equal. Some are built to serve you. Most are built to serve their manufacturers.

Look for devices with:

- Unlocked bootloaders.
- Open-source firmware or support for it.
- Transparent, auditable security practices.

Recommended Devices:

- **Google Pixel 6, 6a, 7, 7a, 8, 8 Pro** – The gold standard for GrapheneOS.
- **Librem 5** – Open-source Linux phone with hardware kill switches.
- **Fairphone 4** – Modular, repairable, privacy-conscious.
- **Unihertz Titan** – Rugged, keyboard-equipped, with user control in mind.

Why Not iPhones? Apple's security is impressive. Its control over you is absolute. No custom ROMs. No root access. No real ability to remove proprietary services. If you need *control*, look elsewhere.

Important: Always purchase new, sealed devices from trusted sources.

- Avoid second-hand or pre-configured devices for sensitive operations.

- Physically inspect packaging for tamper-evidence.

- Re-flash firmware and verify software integrity before first use.

5.3 Installing GrapheneOS

GrapheneOS is more than an Android fork—it's a fortress. Hardened memory management, secure app isolation, no Google surveillance by default.

Step 1: Enable Developer Mode and Unlock the Bootloader

```
1   # Enable Developer Mode
2   Go to Settings > About Phone > Tap "Build Number" 7 times
3
4   # Enable OEM Unlocking
5   Go to Developer Options > Toggle "OEM Unlocking"
6
7   # Reboot into Fastboot Mode
8   adb reboot bootloader
9
10  # Unlock the Bootloader (WARNING: This will wipe all data)
11  fastboot flashing unlock
```

Remember: Unlocking the bootloader is destructive. Back up nothing you plan to keep.

Step 2: Flash GrapheneOS

Download the official images from https://grapheneos.o
rg/releases and flash them:

```
1   cd grapheneos
2   fastboot flash bootloader bootloader.img
3   fastboot flash radio radio.img
4   fastboot reboot-bootloader
5   # Flash OS Components
6   fastboot flash --disable-verity --disable-verification boot
        boot.img
7   fastboot flash system system.img
8   fastboot erase userdata
9   fastboot flashing lock
```

Lock your bootloader again. Always. An unlocked
bootloader is a welcome mat for attackers.

5.4 Other OS Options (When GrapheneOS Isn't an Option)

- **CalyxOS** - User-friendly, includes microG for
  limited Google compatibility.

- **DivestOS** - A hardened fork of LineageOS,
  enhanced with privacy-first modifications.
  While it remains a solid choice, development
  officially ceased in December 2024. It may
  still serve you well in the short term, but as
  of this edition, consider evaluating actively
  maintained alternatives for long-term security
  and support.

- **Ubuntu Touch** - Pure Linux, but app support is
  hit-or-miss.

5.5 Post-Installation Hardening

You have the hardware. You have the OS. Now make it airtight.

1. Enable Full-Disk Encryption

```
1   # Verify encryption status
2   adb shell getprop ro.crypto.state
3   # Should return: encrypted
```

2. Disable Wireless Leakage

- Disable Bluetooth and Wi-Fi scanning.

- Turn off NFC. No one needs to tap your phone in public.

3. Set a Strong Lock Screen

- Long passphrase. No 4-digit PINs. No biometrics. Ever.

4. Ruthlessly Limit App Permissions

- Install only from F-Droid, not the Play Store.

- Revoke camera, microphone, and location permissions unless essential.

5. Install a Secure Messenger

- **Signal** – Industry standard.

- **Session** - Decentralized. No phone number required.

- **Briar** - Works over Bluetooth or Wi-Fi with no central servers.

6. Physically Disable or Cover Sensors (Optional for High-Risk Users)

Even with hardened software, hardware sensors remain exploitable. Consider physical countermeasures:

- Cover cameras with adhesive lens covers or electrical tape.

- Block microphones with hardware mutes or by physically disabling internal microphones (advanced users only).

- Use audio jack blockers to prevent microphone access via headphone ports.

- Avoid smart accessories that add additional microphones or sensors.

5.6 Privacy-Respecting Apps You Should Consider

- **Fennec or Tor Browser** - For private browsing.

- **Mullvad VPN** - Anonymous, no-logs VPN.

- **Aegis Authenticator** - 2FA without vendor lock-in.

- **Bitwarden or KeePassXC** - Manage passwords securely.

- **Simple Mobile Tools Suite** - Calendar, Contacts, Files—all open-source.

5.7 eSIM vs Physical SIM: What You Need to Know

Modern smartphones often come with both eSIM and physical SIM capabilities. While eSIMs are convenient, they pose unique privacy and security risks.

Why It Matters:

- eSIM profiles can be **remotely provisioned or modified** by carriers or governments without physical access.

- Physical SIM cards provide **offline control**—you choose when they are inserted, removed, or destroyed.

- Both SIM types carry unique identifiers (IMSI) that can be tracked across networks and locations.

Recommendations:

- Prefer **physical SIM cards** when operational control is a priority.

- Acquire SIMs using **cash purchases**, avoiding personal information.

- Use **burner SIMs** for compartmentalized identities or travel.

- Rotate SIMs and never reuse the same one across different threat models.

5.8 Long-Term OPSEC Practices

- Never reinstall Google Play Services.

- Use a separate phone for sensitive operations.

- Update regularly—your fortress rots without maintenance.

- Avoid app clutter. Every app is another attack surface.

- Physically protect your microphone and camera. Trust no sensor.

- Use a Faraday bag when traveling to prevent tracking, remote exploitation, or IMSI catcher interception.

- Another reason to store your device in a **Faraday bag** when not in use to block cellular, Wi-Fi, Bluetooth, GPS, and RFID signals—protecting against remote activation or tracking.

- Avoid reusing the same **device IMEI/MEID** across different operational identities or regions. Each phone broadcasts its unique identifier to mobile networks, making it a powerful tracking vector.

Trust no sensor. Treat every microphone, camera, GPS chip, and motion sensor as potential surveillance tools—because they are.

And remember: *Privacy is not a product you install once. It's a practice you live every day.*

6.

Hardware Security

They say you can't secure what you don't control. *What they don't tell you is... you probably don't control your hardware.*

You can install the most hardened OS, lock down every port, and encrypt every byte of data—but if the hardware beneath your fingers is compromised, it's all a performance for an audience you didn't invite.

From firmware backdoors to supply chain sabotage, hardware is the security industry's elephant in the room. It's time we talk about it.

6.1 BIOS & Firmware Backdoors

Your BIOS or UEFI is the first thing your machine trusts. Attackers love that.

Common Firmware-Level Threats:

- **Firmware Backdoors** - Hidden management engines, debug ports, and remote access "features" you never asked for.

- **BIOS Malware (LoJax, MosaicRegressor)** - Stealthy, persistent, and survives OS reinstalls like a cockroach after a nuclear blast.
- **Supply Chain Tampering** - Sometimes the backdoor is pre-installed, straight from the factory.

Countermeasures You Should Be Using:

- **Enable Secure Boot** - Prevents unsigned firmware from running. Not perfect, but better than nothing.
- **Use Open Firmware (coreboot, Libreboot)** - Transparency is power.
- **Update Your BIOS/UEFI Regularly** - Yes, even that scary update from the vendor you don't trust.
- **Neutralize Intel ME / AMD PSP** - These closed-source management engines are black boxes. Consider using me_cleaner to disable Intel ME if your hardware allows it.
- **Verify Firmware Integrity** - Tools like Flashrom can dump and compare your firmware against known good images. Think of it as checking your locks—digitally.

6.2 The TPM Dilemma: Trusted, But Dangerous

The Trusted Platform Module (TPM). Loved by enterprise. Feared by privacy advocates.

Marketed as a security hero, TPMs offer hardware-backed encryption and secure boot features. But in the wrong hands? They're surveillance tools with a silicon smile.

Why TPMs Make Privacy Advocates Nervous:

- **Unique Device Fingerprinting** - TPMs provide hardware IDs. Perfect for corporate tracking or state-level fingerprinting.

- **Remote Attestation** - TPMs can prove the state of your system to a remote party—sometimes without your knowledge.

- **Closed-Source Implementations** - What's in that chip? You'll never really know.

How to Reduce TPM Risks:

- **Disable TPM if You Don't Need It** - If your encryption doesn't require it, kill it.

- **Use Open Verification Tools** - Purism's **PureBoot** offers tamper-evident boot without secretive TPM dependencies.

- **Manual Encryption Key Handling** - LUKS with a strong passphrase keeps you in control of your keys—not your hardware vendor.

- **Prefer Software-Based Encryption** - Software like LUKS, VeraCrypt, or GPG put control back in your hands.

TPM is not inherently evil. But like any powerful tool, it can secure you—or betray you. *The difference depends on who holds the keys.*

6.3 Advanced Firmware Attestation and Tamper Detection

You trust your computer when you turn it on. But the real question is: *Should you?*

Most users have no idea if their firmware has been silently replaced, injected with malware, or otherwise tampered with long before their operating system loads. And once compromised at this level, your OS, your apps, even your encryption—they're all theater.

Enter Firmware Attestation.

Attestation is the process of proving your firmware hasn't been silently altered. Think of it as a digital blood test for your hardware. No surprises. No assumptions.

Measured Boot with Heads or coreboot

Heads Firmware An open-source firmware project built on coreboot, Heads provides:

- Tamper-evident boot verification using cryptographic signatures.

- Trusted boot measurement logs to verify firmware integrity.

- Support for external USB security tokens (Librem Key, Nitrokey) for physical verification.

With Heads, if anything changes—firmware, kernel, boot config—you'll know before your system boots.

How Physical Verification Works

Example: Using a Librem Key, you can physically confirm that your boot process hasn't been tampered with. If the key's LED lights up green, your firmware is clean. *If it turns red? Pull the plug.*

Recommended Tools and Projects

- **Heads** - https://osresearch.net

- **coreboot** - https://coreboot.org

- **Librem Key (Purism)** - Physical USB token for tamper verification.

- **Nitrokey Pro 2 / Nitrokey Start** - Affordable GPG and verification tokens.

These aren't toys. They are tools for those who treat security as a practice, not a product.

If You Can't Install Heads or coreboot

- Regularly snapshot your firmware with flashrom.

- Compare future snapshots to detect unauthorized changes.

- Physically secure your device when not in use to prevent "Evil Maid" attacks.

Remember: A clean boot is the foundation of all security. *Never assume yours is.*

6.4 Supply Chain Verification: Trust No Box

You unbox that shiny new laptop. Factory sealed. Fresh plastic smell. Safe, right?

Wrong.

Between the manufacturer, the shipper, the distributor, and the delivery driver who dropped it on your porch—*how many hands touched that device?* How many could have tampered with it?

Understanding Supply Chain Risk

Supply chain attacks don't require malware. They happen long before you power on. Modified hardware. Pre-loaded backdoors. Tampered firmware.

These attacks are silent. Invisible. Devastating.

Real-World Examples

- **NSA Supply Chain Interdictions** – Documented operations inserting implants into devices in transit.

- **Supermicro Allegations** – Claims of hardware implants in server motherboards.

- **Pre-Compromised Laptops** – Resellers flashing modified firmware before resale.

Anti-Interdiction Strategies

1. Buy Direct from Trusted Vendors

- Avoid third-party resellers.

- Prefer privacy-focused vendors like Purism, System76, Nitrokey.

2. Use Tamper-Evident Packaging and Seals

- Trusted vendors provide tamper seals or void stickers.

- Document packaging on arrival—photos, videos, serial numbers.

3. **Validate Firmware and Software**

- Re-flash firmware from official sources before first use.

- Install your own OS—never trust pre-installed operating systems.

4. **Request Anti-Interdiction Services**

- Vendors like Purism offer custom anti-interdiction:
  - Tamper-evident tape.
  - Photographic proof of device state before shipment.
  - Out-of-band communication to confirm device integrity.

What to Do if You Suspect Compromise

- Contact the vendor immediately.

- Document all evidence.

- Consider the device burned—replace it if your threat model demands certainty.

Trust is earned. Packaging is marketing. *Verify everything.*

6.5 Chipset Isolation: Outsmarting Silicon Spies

Imagine buying a high-security safe—only to find a tiny, locked box inside, labeled "Property of the Manufacturer." That box is the Intel Management Engine (IME). Or, if you swing AMD, the Platform Security Processor (PSP).

The Silicon You Didn't Ask For

Hidden inside nearly every modern Intel and AMD CPU is an embedded microcontroller. It runs its own operating system. It has its own network stack. It operates below your OS, below your BIOS—*below your control*.

Capabilities Include:

- Full memory access.

- Network communication—even when your CPU appears idle.

- Remote administration features designed for enterprises… and sometimes, for nation-states.

You can't inspect the code. You can't audit the firmware. *And you can't easily disable it—by design.*

Why You Should Care

- If compromised, these microcontrollers can bypass all OS-level security.

- They are invisible to standard security tools—no antivirus sees them.

- Supply chain attackers love them because they're always on and rarely monitored.

Neutralizing Intel ME (If You Can)

For select Intel CPUs (usually older generations), you can partially disable Intel ME using tools like me_cleaner.

- **What It Does:** Disables unnecessary modules in Intel ME, leaving just enough to boot the system.
- **Limitations:** Not foolproof. Newer generations are increasingly resistant to neutering.

Learn More: https://github.com/corna/me_cleaner

Libreboot and Coreboot: Minimal Firmware, Maximum Control

If your hardware supports it, replacing your proprietary BIOS with **coreboot** or **Libreboot** is one of the strongest moves you can make.

- Removes vendor backdoors.
- Reclaims control over boot processes.
- Supports verifiable, open-source firmware.

Downside? *Hardware compatibility is limited.* Research before you buy.

Low-Risk Hardware Alternatives

If you can't neutralize Intel ME or AMD PSP, consider switching architectures:

- **ARM-Based Systems** – Many ARM boards lack vendor-locked management engines.

- **RISC-V (Emerging)** – Fully open hardware designs, still maturing but promising.

- **Verified Open Hardware Vendors** – Purism, Nitrokey, and System76 are pushing back against vendor lock-in.

Key Takeaway: If your security starts at the silicon layer, *make sure the silicon works for you—not against you.*

6.6 External Hardware Security Modules (HSMs)

If the phrase "trusted computing" makes you uneasy, good. You should be wary of trusting anything you didn't lock down yourself.

Enter **Hardware Security Modules (HSMs)**—tiny, purpose-built devices that put cryptographic power back in *your* hands, not the manufacturer's.

What Is a Hardware Security Module?

An HSM is a physical device designed to:

- Store cryptographic keys securely.

- Perform authentication without exposing secrets.

- Protect against phishing, malware, and key extraction.

Unlike a TPM soldered to your motherboard, HSMs are:

- **Portable** – You control where and when they're connected.

- **User-Owned** – No manufacturer backdoors (assuming open hardware).

- **Interchangeable** – You can rotate them like you rotate keys.

Recommended Hardware Security Tokens

1. YubiKey (Yubico)

- Industry standard for FIDO2, U2F, OTP, and PGP key storage.

- Supports passwordless login, SSH authentication, and more.

- Proprietary firmware, but widely used and audited.

2. Nitrokey (Pro, Start, HSM)

- Open-source hardware and firmware.

- GPG, SSH, and password storage.

- Enterprise-ready HSM models available.

3. OnlyKey (CryptoTrust)

- Multi-factor authentication with PIN protection.

- Encrypted backup and self-destruct capability.

- Open-source firmware.

What You Can Do With a Security Token

- **Secure SSH Logins** - No more static passwords or exposed keys.

- **Sign Git Commits** - Cryptographically prove your identity.

- **Encrypt and Decrypt Emails** - Secure PGP operations without risking key theft.

- **Two-Factor Authentication (2FA)** - Phishing-resistant login for major platforms.

Why This Matters More Than TPMs

- Tokens are user-controlled, removable, and replaceable.

- No silent remote attestation—*you decide when and where they're used.*

- They complement, not replace, software-based encryption like LUKS or VeraCrypt.

Pro Tip: Always keep a backup token, stored securely in a separate location. One lost token shouldn't lock you out forever.

Control your keys. Control your security.

6.7 RF Isolation and Electromagnetic Threats

What happens in your device… doesn't always stay in your device.

Your laptop, phone, even your cables—they all emit signals. Radio waves, electromagnetic noise, wireless chatter. All of it can be intercepted, analyzed, and—if you're careless—exploited.

The Forgotten Layer: Physical Emissions

This isn't spy fiction. It's science. **TEMPEST attacks**—the art of capturing electromagnetic leaks—are real.

- **Keystroke Logging via RF** - Intercepted from afar.

- **Screen Content Leakage** - Extracted through poorly shielded video cables.

- **Unintentional Bluetooth/Wi-Fi Leakage** - Even when devices appear idle.

Faraday Isolation: The Digital Cloak of Invisibility

A **Faraday Cage** blocks electromagnetic signals from entering or leaving a space. A **Faraday Bag** does the same—for your devices.

What You Can Shield:

- Smartphones.

- Laptops.

- External drives and USB sticks.

- RFID cards and keyfobs.

When to Shield:

- During high-risk travel.

- When storing sensitive devices.

- During private or classified conversations.

Recommended Products:

- **Silent Pocket Faraday Bags** – Trusted by journalists and security professionals.

- **Mission Darkness Shielded Gear** – RF shielding for laptops, phones, and tablets.

- **DIY Shielding** – Copper mesh, aluminum tape, or layered Mylar bags.

Active RF Monitoring

Want to take it further? Monitor your own environment.

- **Software-Defined Radio (SDR)** – Scan for rogue wireless signals with tools like rtl_sdr or HackRF.

- **Bluetooth Sniffers** – Detect hidden Bluetooth beacons or trackers.

- **Wi-Fi Scanners** – Identify unauthorized networks and devices nearby.

Pro Tip: Your phone's "Airplane Mode" isn't perfect. Physical isolation beats software switches every time.

Electromagnetic Hygiene: Your New Habit

- Store idle devices in Faraday bags or shielded rooms.

- Avoid public USB chargers—use charge-only cables or your own power banks.

- Monitor for rogue signals in sensitive environments.

If they can't reach your signals, *they can't reach you*.

6.8 Anti-Interdiction & Tamper-Proofing Techniques

What's the easiest way to compromise a device? Intercept it before it ever reaches the buyer.

It's called **interdiction**—the silent theft or manipulation of hardware during shipping or storage. And unless you're paying attention, you'll never know it happened.

How Interdiction Works

Your laptop ships from a vendor. But it never arrives directly to you.

- It's diverted, unboxed, tampered with.

- Firmware or hardware implants are added.

- It's re-sealed, re-boxed, and sent on its way—looking perfectly untouched.

Congratulations. You just plugged in a compromised machine.

Recognizing High-Risk Scenarios

- Purchasing from large online marketplaces (eBay, Amazon Resellers).

- Ordering high-security devices across international borders.

- Receiving shipments with broken or replaced seals.

- Receiving a device that takes longer than expected to arrive.

Vendor-Supported Anti-Interdiction Services

Some privacy-first vendors offer anti-interdiction protection:

- **Purism Anti-Interdiction:**

    - Tamper-evident tape and unique markings.
    - Out-of-band communication with photos of your sealed device.
    - Optional customizations to detect tampering.

- **System76 Tamper Seals:**

    - Sealed packages with tracking from factory to customer.

DIY Anti-Interdiction Measures

- **Photograph Serial Numbers** – Document packaging, labels, and device markings on arrival.

- **Inspect Packaging Integrity** – Look for tampered seals, re-taping, or non-original materials.

- **Re-Flash Firmware** – Wipe and reinstall trusted firmware or BIOS immediately.

- **Reinstall Your OS** – Never trust factory-installed operating systems.

What to Do If You Suspect Tampering

- Stop. Do not power on the device.

- Photograph all packaging and the device itself.

- Contact the vendor immediately.

- Consider securely destroying the device if your threat model demands certainty.

Remember: If the box looks perfect, assume nothing. *Trust is earned— packaging is just theater.*

7.

Building an Air-Gapped System

An air-gapped system is a computer that is physically isolated from external networks, including the internet and local networks. This isolation prevents remote cyberattacks, malware infections, and unauthorized data exfiltration.

Air-gapped systems are essential for:

- Secure data storage – Protecting sensitive documents, encryption keys, and classified files.

- Cryptographic key management – Storing PGP, GPG, or Bitcoin private keys offline.

- Malware analysis and forensic research – Ensuring that malware cannot connect to a command-and-control server.

- Classified or government operations – Preventing espionage and unauthorized access.

Choosing the Right Hardware

A proper air-gapped system should have no networking hardware (Wi-Fi, Bluetooth, Ethernet) and must allow full user control over firmware and software.

Recommended Devices for Air-Gapped Use:

- ThinkPad X230, T480 (Libreboot/Coreboot) — Best for installing open-source firmware.

- Purism Librem 14 — Features hardware kill switches to disable Wi-Fi, microphone, and camera.

- Raspberry Pi (Minimal footprint) — A small, low-power offline system for isolated tasks.

- Custom-Built Offline Desktops — No networking hardware, full BIOS/UEFI control.

Preparing the Device for Air-Gapped Use

Before using the system, follow these critical steps:

1. Physically Remove Networking Hardware — Disconnect Wi-Fi cards, Ethernet ports, and Bluetooth modules.

2. Install an Open-Source BIOS (If Possible)

   ```
   # Flash Libreboot or Coreboot
   flashrom -p internal -w libreboot.rom
   ```

3. Disable USB Autorun — Prevent malware from executing automatically via USB.

4. Use a Minimalist, Security-Focused OS — Recommended: Qubes OS, Tails OS, or a hardened Linux distro.

Setting Up an Air-Gapped Operating System

The best operating systems for air-gapped use are:

- Qubes OS - Provides security through compartmentalized virtual machines.

- Tails OS - A live system designed for privacy,
  anonymity, and zero data persistence.

Installing Qubes OS:

1. Download the latest Qubes OS ISO from https://ww
   w.qubes-os.org.

2. Verify the PGP signature to ensure authenticity.

3. Flash the image to a USB drive:

```
sudo dd if=qubes.iso of=/dev/sdX bs=4M status=
progress
```

4. Install on a dedicated, non-networked machine.

Installing Tails OS (For Disposable Sessions):

1. Download the latest Tails OS ISO from https://ta
   ils.net.

2. Verify the signature using:

```
gpg --verify tails-signature.asc tails.iso
```

3. Burn the image to a USB stick and boot into a
   fully ephemeral system.

Secure Data Transfer (Data Diodes)

Since air-gapped machines cannot connect to networks,
data transfer requires special precautions to prevent
leaks.

USB Sneakernet (Basic Method)

A sneakernet involves manually moving data between systems using a USB drive or SD card. To do this securely:

- Use a brand new, trusted USB drive - Never reuse a USB that was previously connected to an online system.

- Encrypt all data using VeraCrypt or GnuPG before transferring.

- NEVER plug the same USB into both an air-gapped and networked system.

Encrypting Files with VeraCrypt:

```
1  # Create an encrypted container
2  veracrypt -t -c
3
4  # Mount the encrypted volume
5  veracrypt /dev/sdX1 /mnt/secure
```

One-Way Transfer Using a Data Diode

A data diode ensures one-way data flow, preventing an air-gapped system from accidentally leaking information.

Building a DIY Data Diode with a Raspberry Pi:

1. Set up two Raspberry Pis - one connected to a networked system, the other to the air-gapped system.

2. Use an optical cable to enforce one-way data transmission.

3. The online Pi sends data, while the offline Pi can only read but not send anything back.

QR Code Data Transfer (For Small Files)

For small, sensitive files (such as Bitcoin wallets or encryption keys), use QR codes for offline transfer.

```
1  # Convert file to a QR code
2  qrencode -o secure.png -s 10 < secretfile.txt
3  # Scan QR code on air-gapped system
4  zbarimg secure.png
```

Preventing Side-Channel Attacks

Even air-gapped systems can be compromised through side-channel attacks, which extract information through unintended leaks such as electromagnetic radiation or acoustic signals.

Attack Type	How It Works	Defense
TEMPEST	Captures electromagnetic emissions from a device	Use a Faraday cage, shielded cables
AirHopper	Uses GPU to transmit data via radio waves	Disable speakers, use audio jammers
Cold Boot	Extracts encryption keys from RAM	Power off system completely before removal
Acoustic Leakage	Captures keystrokes via sound waves	Use white noise generators

Table 7.1: Common Air-Gap Attack Methods and Countermeasures

Final OPSEC Recommendations

To ensure maximum security, follow these best practices:

- Store the air-gapped system in a Faraday cage when not in use.

- Disable microphones, cameras, and all network interfaces.

- NEVER insert untrusted USB devices into the system.

- Assume all electromagnetic emissions are vulnerable to eavesdropping.

If you truly need to lock something down—if you truly need to build a digital black site—there's only one move left:

Total Isolation.

No Wi-Fi. No Bluetooth. No Ethernet. No signals. No connections. A fortress sealed from the world, immune to remote compromise.

In the next chapter, you'll build one. **An Air-Gapped System.**

Welcome to true operational security.

7.1 Going Beyond Isolation: Real-World Air-Gap Threats

Building an air-gapped system is a powerful move. But isolation alone isn't immunity.

Advanced attackers have developed creative ways to *bridge* air gaps—literally jumping data across the void using techniques you wouldn't believe if they weren't published in peer-reviewed research.

Gap-Jumpers: Attacks That Cross the Air Gap

Air-gapped doesn't mean air-tight. Here are some of the most shocking methods used to exfiltrate data from air-gapped systems:

- **AirHopper** – Uses GPU-induced radio waves to transmit data to nearby smartphones.

- **PowerHammer** – Modulates CPU load to send data through power lines.

- **GSMem** – Uses electromagnetic interference near cellular bands to broadcast data.

- **Fansmitter** – Varies CPU fan speeds to encode data as acoustic signals.

- **LED-it-Go** – Flashes status LEDs at imperceptible rates to encode binary data.

Realistically, Should You Be Worried? If you're handling national security data or corporate trade secrets—*yes*. For general personal privacy use—probably not, but awareness costs nothing.

Mitigation Checklist:

- Disable or limit CPU fan control in firmware settings.

- Physically cover or disable unnecessary LEDs.

- Remove or disable speakers and audio outputs.

• Store systems in RF-shielded enclosures if handling high-value data.

Enforcing One-Way Data Policies

Most air-gap failures happen not through technical wizardry, but through careless data transfers.
You need a Data Handling Policy:

- **Ingress Devices (Data In):** Dedicated USB drives or SD cards that only ever bring data *into* the air-gapped machine. Once used, they **never** go back online.

- **Egress Devices (Data Out):** Separate, clean USB drives used only to *export* data. These devices are scanned, validated, and handled under strict supervision before ever connecting to a networked machine.

Best Practices for Data Transfers:

- Use new, factory-sealed storage media.

- Encrypt all data before transferring it in or out.

- Verify file integrity using hashes or GPG signatures before release.

- Sanitize or physically destroy media after one-time use.

Operator Discipline: The Human Air Gap

You can air-gap your hardware, but if you don't air-gap your behavior, you're still exposed.

Behavioral Discipline Includes:

- Leaving phones, smartwatches, and RF-enabled devices outside the air-gapped environment.

- Using separate clothing, gloves, or static-free garments in ultra-sensitive scenarios.

- Ensuring only trusted personnel interact with the air-gapped system.

Environmental and Insider Risks (Awareness Level)

While most readers won't have access to shielded rooms or government-grade isolation, here's what to know:

- **RF-Shielded Rooms:** Expensive, but block all wireless signals.

- **EMI Shielded Cabling:** Prevents data leakage through electromagnetic emissions.

- **Physical Supervision:** Human oversight is still your best defense against insiders.

These measures are more relevant for military, intelligence, or critical infrastructure environments—but knowing they exist helps frame your own security model.

Reality Check

Most air-gap failures are human, not technical. Fail to control media. Fail to control behavior. Fail to control who has access.

The air gap is only as strong as the discipline behind it.

Advanced Side-Channel Attacks on Air-Gapped Systems

Air-gapped systems are vulnerable to physical emissions that leak data without needing network access.

Examples of Side-Channel Attacks:

- **Acoustic Emissions:** Data leaked via fan speed or hard drive noise (e.g., Fansmitter, DiskFiltration).

- **Optical Signals:** Data exfiltrated through blinking LEDs (keyboard, router) or screen brightness.

- **Magnetic or Power Line Emissions:** Monitoring electromagnetic or power fluctuations to extract data.

Countermeasures:

- Shield systems using **Faraday enclosures.**

- Disable non-essential hardware (LEDs, speakers, fans).

- Monitor for unauthorized emissions with EMF scanners.

7.2 Building a Real-World Air-Gapped Workstation

Talk is cheap. Let's build one.

This section provides a practical, repeatable blueprint you can follow to build a fully offline, air-gapped workstation—from hardware selection to environmental setup.

Required Materials

- **Hardware:** A laptop or desktop with user-replaceable parts. Example: ThinkPad T480 or a custom desktop build.

- **Tools:** Precision screwdriver set, anti-static wrist strap, thermal paste (if reassembling CPUs).

- **Software:** - Coreboot/Libreboot (optional but recommended). - Qubes OS or Minimal Linux Distro ISO (verified with PGP). - Flashing tools (flashrom, USB ISO writer).

- **Security Materials:** - Epoxy or USB port blockers (optional but recommended). - Tamper-evident seals or labels. - Faraday bag or shielded storage container.

Step-by-Step Physical Build Process

Step 1: Disassemble the Device

- Remove the back panel.

- Locate and physically remove:

  - Wi-Fi card.
  - Bluetooth modules.
  - Cellular modem (if installed).

- Any unnecessary antennas.

Step 2: Secure BIOS/UEFI Settings

- Boot into BIOS/UEFI setup.

- Disable:

  - PXE (Network Boot).
  - Bluetooth and Wi-Fi (if still present).
  - Audio devices (optional).
  - USB boot (optional).

- Set a strong BIOS administrator password.

Step 3: Install Open-Source Firmware (Optional but Recommended)

- Flash Coreboot or Libreboot if supported:

```
flashrom -p internal -w libreboot.rom
```

- Reboot and verify functionality.

Step 4: Install a Minimal, Offline OS

- Flash Qubes OS or your preferred hardened Linux ISO to a USB drive.

- Boot and verify the PGP signature.

- Install the OS with:

  - No networking packages.
  - No wireless drivers.

Step 5: Physically Lock Down Ports (Optional)

- Apply epoxy or physical port blockers to:

- Ethernet ports.
- Unused USB ports.
- Audio jacks (if acoustic attacks are a concern).

Step 6: Apply Tamper-Evident Seals

• Place seals on case seams, screws, and ports.

• Photograph and document serial numbers and seal positions.

Step 7: Environmental Hardening

• Store in a Faraday bag or shielded box when not in use.

• Prohibit RF-enabled devices in the room during operation.

• Physically secure the device in a locked cabinet or safe.

What You've Built

• No Wi-Fi. No Bluetooth. No Ethernet.

• Minimalist OS with no external communication stack.

• Physically and logically locked down.

• Ready for high-assurance, offline operations.

This is how you build a true air-gapped workstation. **Not unplugged. Not hidden. Physically severed from the world.**

7.3 Closing Thoughts on Hardware Security

You've learned to challenge the sanctity of firmware, scrutinize supply chains, strip silicon of its silent spies, and defend against signals you can't even see.

But here's the bitter truth: *Every electronic device you own is a potential liability.*

Your laptop, your phone, your desktop—they're all loudmouths in the right electromagnetic spectrum. They leak data. They beg to connect. They're built to share.

7.4 Final Reality Check: Is Air-Gapping Right for You?

Let's be clear—air-gapped systems are **not** everyday computers. They are tools for very specific, high-assurance tasks.

If you just want to block ads, avoid tracking, or browse privately—this isn't the move. Air-gapped systems are inconvenient by design. No network. No cloud. No convenience.

But if you need a digital vault— If you manage cryptocurrency keys, sensitive research, or operational data worth stealing— If you'd rather destroy a USB stick than let it touch the wrong machine— *This is how you build that vault.*

And remember, no machine is secure if the human using it isn't.

Air-gapping isn't a setting. It's a discipline. Make sure you're ready to practice it.

By following these methods, you will have a fully

secure, air-gapped system resistant to remote attacks and data leaks.

8.

Secure Web Browsing & Anonymous Internet Use

The internet was supposed to set us free. Instead, it became the most invasive surveillance system in human history.

Every click, every search, every purchase—it's all recorded, categorized, sold, and weaponized. Not just by corporations. Not just by governments. But by anyone with the resources to buy your behavioral data wholesale.

Cookies? That's child's play. Your browser leaks far more than you realize.

8.1 Why Browsers Are Built to Betray You

Your browser is the front door to your digital life—and most doors come with peepholes facing both ways.

- **Cookies and Supercookies** - Tracking you across sites.

- **Fingerprinting** - Profiling your hardware,

software, and behavior.

- **WebRTC and DNS Leaks** - Broadcasting your real IP address without warning.

- **Telemetry** - "Performance data" you never agreed to share.

Want to fight back? You'll need better tools, sharper habits, and the discipline to use both.

8.2 Choosing a Browser That Works For You (Not Against You)

Here's where most people get it wrong: **Privacy isn't what you install. It's what you stop trusting.**

Forget Chrome. Forget Edge. You need a browser that fights *for* your privacy, not against it.

- **Firefox (Hardened)** - Open-source, highly customizable.

- **LibreWolf** - Firefox on privacy steroids, no telemetry, hardened defaults.

- **Tor Browser** - The anonymity gold standard, but slow by design.

- **Mullvad Browser** - Anti-fingerprinting by default, built for VPN or Tor users.

Harden Firefox Like You Mean It:

- Disable WebRTC, geolocation, and telemetry in about:config.

- Install CanvasBlocker and Chameleon.

- Enable privacy.resistFingerprinting = true.

8.3 Psychological Manipulation: Defeating Dark Patterns

You've blocked the cookies. You've masked the fingerprint. You've routed your traffic through Tor or a VPN.

And yet—you still lose.

Not because of your browser, but because of your **behavior**. Welcome to the world of **dark patterns**—web design tricks engineered to make you betray yourself.

What Are Dark Patterns?

Dark patterns are deceptive design techniques used to manipulate your decisions, typically by:

- Nudging you to accept tracking "for your convenience."

- Burying privacy settings behind confusing menus.

- Using color, size, and placement to steer you toward bad choices.

- Guilt-tripping or scaring you into clicking "Agree."

You've seen them. Those bright green "Accept All Cookies" buttons, with the tiny gray "Manage Settings" link almost invisible by design. Those consent forms that make rejecting tracking *ten times harder* than accepting it.

Common Dark Pattern Tactics

- **Forced Consent** – Making you click "Accept" to view content, even when alternatives exist.

- **Hidden Opt-Outs** – Burying privacy settings behind multiple confusing pages.

- **Pre-Checked Boxes** – Opting you in by default, betting you won't notice.

- **Guilt Language** – "Help us improve your experience" or "Support free content by accepting cookies."

- **Deceptive Button Design** – Making "Accept" big, colorful, and prominent, while "Reject" is small or hidden.

Why Dark Patterns Work

These techniques exploit human psychology:

- **Decision Fatigue** – You're tired. You just want to read the article.

- **Social Engineering** – The site uses emotional language to pressure you.

- **Default Bias** – You assume the big green button is the "correct" choice.

They turn your *own mind* into the weakest link.

How to Fight Back

1. Preempt Consent Banners Entirely

- Use **uBlock Origin** with filter lists like "EasyList Cookie Notices" or "I Don't Care About Cookies" to block banners before they load.

2. **Always Look for Manual Settings**

- Avoid "Accept All" buttons.

- Click through to "Manage Settings" or "Reject All" even if it takes longer.

3. **Use CSS/JavaScript Blockers**

- Extensions like **uMatrix** or **NoScript** can block manipulative scripts entirely—just beware of breaking site functionality.

4. **Understand It's Designed to Exhaust You**

- Take a breath. Remember they *want* you to get lazy.

- Train yourself to spot manipulation—*and don't play their game.*

The Real Power Move

Skip the consent banners entirely. Use tools that block trackers before they ever load. Don't wait to be asked for your privacy—you already own it.

Privacy is not a setting. It's the courage to say *No*—even when they make it hard.

Operational Browser Compartmentalization

You wouldn't use the same password everywhere— So why use the same browser for everything?

Browsers are stateful memory machines. They remember what you do, what you log into, what you search. **Even the best-hardened browser can betray you if you mix personal and private activities.**

The Case for Separation

Different tasks deserve different environments:

- **Personal Accounts** – Google, banking, social media.

- **Anonymous Research** – OSINT, privacy research, whistleblowing.

- **Operational Security** – High-risk tasks, dark web access, sensitive communications.

- **Disposable Sessions** – One-time lookups, burner identities, untrusted links.

Example Browser Isolation Strategy

Browser	Purpose
Firefox (Profile 1)	Personal accounts, non-sensitive browsing.
LibreWolf or Hardened Firefox (Profile 2)	Privacy-first browsing, research without login.
Tor Browser	Anonymous browsing, dark web access, high-risk operations.
Mullvad Browser	Disposable sessions, OSINT, one-time tasks.

Table 8.1: Example Browser Compartmentalization Strategy

Pro Tips for Compartmentalization

- Never cross streams—*don't log into personal accounts from your privacy browsers.*

- Clear cookies and site data regularly.

- Use different VPN exit nodes or Tor identities for separate tasks.

- Consider using separate physical devices for high-risk operations.

Isolation is a habit. Start practicing it before you actually need it.

8.4 Why Fingerprinting Matters More Than Cookies

Tracking pixels and cookies are noisy but easy to block. Fingerprinting is silent, persistent, and terrifyingly effective.

You Are Fingerprinted By:

- Your screen size.

- Your fonts and language settings.

- Your time zone and hardware acceleration quirks.

- Your GPU, audio stack, and canvas rendering output.

Real Defense:

- Tor Browser (Best-in-class).

- Mullvad Browser (Fingerprint masking).

- Hardened Firefox (Manual but effective).

8.5 Silencing Trackers and Ad Networks

Modern advertising isn't advertising—it's surveillance.

Stop the Signal:

- **uBlock Origin** - Block ads, trackers, and known malware domains.

- **Privacy Badger** - Machine-learned tracker blocking.

- **LocalCDN and Decentraleyes** - Prevent external script-based tracking.

Harden Firefox's Internal Protections:

```
1   # Enhanced Tracking Protection (Strict Mode)
2   privacy.trackingprotection.enabled = true
3
4   # Block third-party cookies
5   network.cookie.cookieBehavior = 1
```

8.6 Mobile Browsing: Your Pocket Is Leaking Too

You've locked down your desktop. You've hardened your browser. You've compartmentalized your online life.

But there's a supercomputer in your pocket… And it's selling you out every time you unlock it.

Mobile Browsers Are Even Worse

Mobile browsers are limited by design—fewer settings, fewer extensions, and fewer privacy protections. Worse, you're probably using one of the big offenders:

- **Google Chrome** – Built by the same company profiting from your data.

- **Safari** – Better than Chrome, but still Apple's data faucet.

- **Default Samsung/Manufacturer Browsers** – Just don't.

You can do better. Switch to privacy-focused alternatives built for mobile defense.

Recommended Privacy Browsers for Mobile

- **Firefox Focus (iOS/Android)** - Minimalist, auto-clears history on exit. - Great for disposable, single-use searches.

- **DuckDuckGo Privacy Browser (iOS/Android)** - Blocks trackers and enforces HTTPS. - Burn button clears all tabs and data in one tap.

- **Brave Browser (iOS/Android)** - Built-in ad and tracker blocking. - Optional Tor tab feature (Android only).

- **Mullvad Browser (Android via .apk)** - Advanced anti-fingerprinting, designed for VPN/Tor users. - No telemetry, no tracking.

- **Bromite (Android only)** - Hardened Chromium fork with built-in ad-blocking. - Privacy-friendly without needing extensions.

Mobile Browsing Hygiene Tips

- **Clear Tabs and History Regularly** Most mobile browsers have a "burn" or "clear" button—use it.

- **Avoid Logging Into Personal Accounts on Privacy Browsers** Compartmentalization applies on mobile, too.

- **Block Trackers at the Network Level** Use apps like NetGuard or Blokada (Android) to block unwanted traffic across all apps.

- **Disable Unnecessary Permissions** Location, microphone, camera—turn them off unless absolutely needed.

- **Use a Privacy-Respecting Keyboard** Keyboards like OpenBoard (Android) don't phone home with everything you type.

Pro Tip: Device-Level Defense

Want to go nuclear? Use a de-Googled Android phone (GrapheneOS, CalyxOS) with zero Google services. No Play Store, no tracking, no telemetry—just control.

Because privacy isn't just for desktops. Your phone deserves better, too.

8.7 DNS Leaks: Your ISP's Favorite Snack

Your ISP sees every domain you visit—unless you stop them.

Block ISP Surveillance with:

- DNS-over-HTTPS (DoH).

- Privacy-first DNS providers like Quad9 or NextDNS.

- Network-level blocking with Pi-hole.

Example: Enabling DoH in Firefox

```
1    network.trr.mode = 3
2    network.trr.uri = "https://dns.nextdns.io"
```

Example: Blocking Ads Network-Wide with Pi-hole

```
1    # Install Pi-hole
2    curl -sSL https://install.pi-hole.net | bash
3
4    # Set to privacy-respecting DNS
5    pihole -a setdns 9.9.9.9
```

8.8 Operational Discipline: The Human Firewall

No browser setting protects you if you hand your data over willingly.

Operational Best Practices:

- Never log into personal accounts (Google, Facebook) from your privacy browser.

- Compartmentalize: One browser for personal, another for private tasks.

- Use Tor or VPNs to hide your IP address.

- Clear cookies and cache after each sensitive session.

- Enable First-Party Isolation in Firefox to block cross-site tracking.

The VPN Illusion: What They Do—and Don't—Protect

You've seen the marketing:

> *"Military-grade encryption." "Hide your location." "Browse anonymously."*

What they don't say? **VPNs aren't magic.** And they don't make you invisible.

What a VPN Actually Does

When you use a VPN, you:

- **Hide your real IP address** from websites and services.

- **Encrypt your internet traffic** between you and the VPN server.

- **Bypass ISP or local network monitoring** (like public Wi-Fi eavesdropping).

- **Circumvent geo-restrictions** (appear to be in another country).

What a VPN Does NOT Do

What VPNs can't protect you from:

- **The VPN Provider Themselves** They can see everything you do through their service—unless they prove otherwise.

- **Browser Fingerprinting and Tracking** You still leak behavioral and device-level data unless you harden your browser.

- **Compromised Devices** If you have malware on your machine, a VPN won't stop it from exfiltrating data.

- **Malicious Websites** VPNs don't block phishing, scams, or malicious content by default.

- **Law Enforcement Requests** If you're logged into personal accounts, your VPN changes nothing.

The VPN Trust Dilemma

You're just moving trust from your ISP to your VPN provider.

Ask yourself:

- Do they log your traffic?
- Are they based in a country with privacy-hostile laws?
- Have they ever been caught lying about "no logs" policies?
- Do they accept anonymous payment methods (crypto, cash)?

Recommended VPN Qualities

- **Strict No-Logs Policy** - Verified by independent audits.
- **Outside 5/9/14 Eyes Jurisdictions** - Prefer privacy-friendly legal environments.
- **Accepts Anonymous Payment** - Crypto or cash.
- **No Bandwidth or Connection Limits** - Avoid "free" VPNs—they pay themselves by logging and selling your data.
- **Transparent Ownership** - Avoid providers who hide behind shell companies.

Trusted VPN Examples (As of Writing)

- **Mullvad VPN** - No email required, accepts cash, Swedish-based with proven no-logs.
- **IVPN** - Anonymous accounts, strong privacy policies.
- **ProtonVPN** - Swiss-based, open-source clients, trusted reputation.

VPN + Tor: The Layered Defense

For maximum anonymity:

- Use a VPN to hide Tor usage from your ISP.

- Or use Tor *without* a VPN to avoid trusting a VPN provider at all.

Warning: VPN-over-Tor and Tor-over-VPN are different—know your tools before you combine them.

The Real Takeaway

A VPN is **just a tool**. *Your privacy depends on how you use it— not on the marketing hype you bought it from.*

Remember: Privacy isn't a browser extension. It's the decision to stop giving your life away—*one click at a time.*

9.

Secure File Storage & Encryption

Data stored on a device is vulnerable to theft, malware, unauthorized access, and forensic recovery. Encryption ensures that even if your files are stolen, they remain unreadable without the decryption key.

Core encryption strategies include:

- Full-Disk Encryption (FDE) - Encrypts the entire storage device.

- Encrypted Containers - Protects selected files inside a secure volume.

- Secure File Sharing - Transfers encrypted files safely.

- Backup Encryption - Ensures data remains protected even in offsite storage.

9.1 Choosing the Right Encryption Strategy for the Right Data

Not all data requires the same level of protection. Here's a practical decision matrix to help you match *what* you're protecting with *how* you protect it.

Data Type	Recommended Method
Personal docs, notes	VeraCrypt container or LUKS encrypted partition
Entire laptop or desktop	Full-Disk Encryption (LUKS, BitLocker*, FileVault)
Cryptocurrency wallets, private keys	Offline, air-gapped storage or Shamir's Secret Sharing
Large backups	Encrypted backup tools like Restic/Borg + LUKS on external drives
Sensitive file sharing	GPG, OnionShare, or Magic Wormhole

Table 9.1: Data Types and Matching Encryption Methods

> **Note**
>
> *BitLocker is okay to use, it is simple and offers a basic layer of protection. However, it does not provide the same level of security, transparency, or auditability as VeraCrypt.

9.2 Defending Against Coercion: Hidden Volumes

Sometimes, being able to open *something* isn't enough—you may be forced to unlock your encrypted data under duress.

Solution: VeraCrypt Hidden Volumes

- **How It Works:** Two passwords unlock two different volumes—one "decoy," one hidden. Observers cannot prove the hidden volume exists.

- **Why It Matters:** Provides plausible deniability if forced to reveal a password under coercion.

- **Best Practice:** Populate the decoy with believable, non-critical data.

9.3 RAM Data Exposure & Cold Boot Attacks

Even with strong encryption, data may temporarily reside in RAM (Random Access Memory) while in use. Cold boot attacks exploit this by physically rebooting a machine and dumping residual memory before it decays.

Threat Model:

- Attackers with physical access freeze the RAM chips to slow data decay.

- They reboot the system to a custom OS and extract decryption keys from memory.

Mitigation:

- Fully power off devices when not in use (shutdown, not sleep or hibernate).

- Use devices with encrypted memory (hardware-dependent).

- Lock screens should trigger memory zeroing where supported (e.g., Tails OS).

- Remove power sources (battery and cable) if leaving devices unattended.

9.4 The Metadata Problem: What Encryption Doesn't Hide

Encryption protects content, not context.
Even fully encrypted volumes may leak:

- File sizes and counts.

- Access timestamps.

- File structure (when using basic compression or unpadded encryption).

Mitigation:

- Use encrypted containers instead of leaving loose files.

- Enable volume size padding when possible.

- Zero-fill free space on encrypted drives to mask usage patterns.

9.5 Secure Deletion: Shredding Unencrypted Data

Deleting a file doesn't erase it. Most file systems simply mark space as "free" without overwriting the contents.

Example: Secure File Deletion on Linux

```
1   # Overwrite a file 5 times before deletion
2   shred -u -n 5 sensitivefile.txt
```

Windows Alternative:

- Use tools like **BleachBit** or built-in **Cipher.exe:**

```
1   # Wipe free space on C: drive
2   cipher /w:C:\
```

Warning: Shredding SSDs is unreliable due to wear-leveling. Use full-disk encryption from the start or secure physical destruction.

9.6 Secure Memory Erasure

Sensitive data in RAM should be securely erased after use to prevent memory scraping attacks.

Threat Model:

- Memory forensics tools may recover passwords or keys if memory is not properly cleared.

Mitigation:

- Use encryption software with **secure memory handling** (e.g., VeraCrypt, GnuPG).

- Prefer tools that call `mlock()` or similar functions to prevent memory swapping.

- Physically power off systems after use to clear volatile memory.

- Avoid leaving decrypted files or keys in RAM longer than necessary.

9.7 The Truth About "Hardware Encrypted" USB Drives

Many cheap "encrypted" USB drives offer fake security:

- Proprietary, unaudited encryption algorithms.

- Weak or nonexistent password protection.

- Vulnerabilities   discovered   in   brand-name products.

Verified Secure Hardware Storage:

- **Apricorn Aegis Secure Key** – PIN-protected, hardware-encrypted storage.

- **YubiKey Storage Modules** – Secure key and file storage for high-assurance use.

Always prefer open-source, peer-reviewed encryption tools over marketing hype.

Full-Disk Encryption (FDE)

Encrypting your entire disk prevents unauthorized access if your device is lost, stolen, or seized. This protects operating system files, swap memory, and user data.

Linux: Encrypt with LUKS (Linux Unified Key Setup)

```
1   # Encrypt a disk with LUKS
2   sudo cryptsetup luksFormat /dev/sdX
3
4   # Open and create a mapped encrypted volume
5   sudo cryptsetup open /dev/sdX encrypted_drive
```

Best Practices for LUKS:

- Use Argon2id key derivation ('cryptsetup luksFormat –type luks2 –pbkdf argon2id').

- Store recovery keys offline (never in plaintext on the same device).

- Use YubiKey or TPM-based unlocking for added security.

Windows: Encrypt with BitLocker or VeraCrypt

- BitLocker (Windows Pro/Enterprise) - Built-in full-disk encryption.
- VeraCrypt (Windows/Linux/macOS) - Open-source encryption alternative.

Enabling BitLocker:

```
# Enable BitLocker with TPM protection
manage-bde -on C: -RecoveryPassword
```

VeraCrypt Full-Disk Encryption:

- Download and install VeraCrypt.
- Select "Encrypt a Partition or Drive".
- Choose AES or Serpent encryption and set a strong passphrase.

Creating Encrypted Containers

Instead of encrypting an entire disk, encrypted containers allow you to protect specific files inside a secure, password-protected volume.

Using VeraCrypt (Cross-Platform):

```
# Create an encrypted container
veracrypt -t -c

# Mount the encrypted volume
veracrypt /dev/sdX1 /mnt/secure
```

Why Use Encrypted Containers?

- Portable – Can be moved between devices.

- Hidden volumes – Protect against coercion attacks.

- Supports AES, Twofish, and Serpent encryption.

Password-Derived Key Strengthening (PBKDF2/Argon2)

Passwords must be converted into cryptographic keys through a Key Derivation Function (KDF). Weak KDFs make brute-force attacks feasible, especially on stolen encrypted volumes.

Recommended KDFs:

- **Argon2id** – Memory-hard and CPU-intensive, highly resistant to cracking.

- **PBKDF2** – Still widely used but less resistant than Argon2id.

Best Practice:

- Always choose encryption software that supports **Argon2id** or allow configuring iteration counts.

- Use high iteration counts or memory parameters (e.g., 1GB+ RAM usage for Argon2).

- Store the KDF configuration alongside encrypted volumes for reproducibility.

Preparing for Quantum Threats

Quantum computers may eventually break RSA, ECC, and other algorithms.

Emerging Solutions:

- **CRYSTALS-Kyber, Falcon, Dilithium** – Post-quantum cryptography candidates.

- Hybrid encryption combining classical and quantum-resistant algorithms.

Recommendation:

- Monitor developments from **NIST Post-Quantum Cryptography Project**.

- Flag long-term sensitive data as **quantum-vulnerable**.

Secure File Sharing

Transmitting files over email, cloud storage, or messaging apps exposes them to interception. Instead, use end-to-end encrypted methods.

Best Encrypted File Transfer Tools:

- OnionShare – Anonymous file sharing over the Tor network.

- Magic Wormhole – Secure file transfer using shortcodes.

- Rclone + Crypt – Encrypts files before syncing them to cloud storage.

Encrypt a file before transferring it:

```
# Encrypt a file using GPG
gpg --output securefile.gpg --encrypt --recipient
    user@example.com file.txt

# Decrypt a received file
gpg --decrypt securefile.gpg > decrypted.txt
```

Protecting Backups

Even backups are vulnerable to data leaks, theft, and
surveillance. To ensure security, backups should be
encrypted, redundant, and stored offsite.

Best Practices for Secure Backups:

- Store backups on air-gapped external drives
  whenever possible.

- Use BorgBackup or Restic for encrypted,
  incremental backups.

- Keep an offsite encrypted backup to prevent data
  loss from theft or disasters.

- Use Shamir's Secret Sharing (SSS) for cryptographic
  key backups.

Using Restic for Encrypted Backups:

```
1   # Initialize a secure backup repository
2   restic init -r /mnt/backup --password-file /root/passphrase.
       txt
3
4   # Create an encrypted backup
5   restic backup /home/user/Documents -r /mnt/backup
```

Encrypting External Drives with LUKS:

```
1   # Format an external USB drive with LUKS encryption
2   sudo cryptsetup luksFormat /dev/sdb1
3   # Open the encrypted USB drive
4   sudo cryptsetup open /dev/sdb1 backup_drive
5
6   # Mount and use the drive securely
7   sudo mount /dev/mapper/backup_drive /mnt/backup
```

9.8 Cloud Disk Snapshot Exploits

In enterprise and cloud environments, snapshots of virtual machines or storage volumes can silently bypass encryption if taken while the system is running.

Threat Model:

- Cloud providers or attackers with administrative access may snapshot encrypted volumes *while they are mounted and unlocked*.

- This bypasses encryption entirely because the data is captured in plaintext.

Mitigation:

- Never trust cloud environments with live decrypted data.

- Always shut down and unmount encrypted volumes when not in active use.

- Choose providers with customer-managed encryption keys (CMEK) or client-side encryption.

- Prefer on-premises encrypted backups over cloud-based live storage.

Final Encryption Recommendations

For maximum security, follow these encryption guidelines:

- Always use full-disk encryption on laptops, desktops, and mobile devices.

- Encrypt external backups and USB drives to prevent physical data theft.

- Use strong passphrases (at least 12+ words) or hardware keys for unlocking.

- Avoid cloud storage without client-side encryption (use Cryptomator or Rclone-Crypt).

- Regularly update encryption software to prevent attacks on outdated algorithms.

By implementing these encryption techniques, your data will remain secure even in cases of physical theft, malware infections, or government surveillance.

9.9 The Real Takeaway

Encryption isn't just a checkbox. It's a strategy. **Choose the right tool for the right threat—and practice securing your keys like your data depends on it.** *Because it does.*

10.

Hardware Attacks & Physical Threats

Even with strong encryption, an attacker with physical access can compromise a system through hardware-based attacks. This section covers the most common threats and the best defenses against them.

10.1 USB-Based Attacks

USB devices are commonly used for malware injection, data theft, and even physical destruction.

Common USB-based threats:

- BadUSB – Reprogrammed USB devices that appear as keyboards or network adapters to execute hidden commands.

- Rubber Ducky – A USB device that emulates a keyboard to inject malicious scripts at high speed.

- USB Kill – Sends a high-voltage electric surge into the USB port, physically frying the motherboard.

- USB Data Exfiltration – Attackers use rogue USBs to steal files from a system automatically.

Important Distinction: USB Kill vs usbkill

There are two tools with nearly identical names but entirely opposite purposes:

- **USBKill (Hardware Attack Device):** A malicious USB stick that delivers high-voltage surges to destroy computer hardware.

- **usbkill (Software Defense Tool):** An anti-forensic Python script that monitors USB ports and immediately shuts down your computer if a new USB device is inserted.

Data blockers ("USB condoms") prevent data theft but do not stop hardware-damaging devices like the USBKill stick. usbkill (the script) is a defensive tool that can help protect against physical attacks by forcing an immediate shutdown when unauthorized USB devices are connected.

How to Defend Against USB Attacks:

- Disable USB autorun to prevent automatic execution of malicious payloads.

- Use USB data blockers (USB condom) when charging devices in public areas.

- Physically disable USB ports on critical machines (epoxy filling or BIOS/UEFI settings).

- Limit USB device permissions (use 'usbguard' on Linux or Device Control on Windows).

- Only use trusted USB devices – Avoid using found USBs or those from untrusted sources.

Linux: Blocking Unauthorized USB Devices

```
1  # Block all new USB devices (Linux)
2  echo 0 | sudo tee /sys/bus/usb/drivers_autoprobe
```

Windows: Disable USB Ports in Group Policy

```
1  # Disable USB storage access (Windows PowerShell)
2  reg add HKLM\SYSTEM\CurrentControlSet\Services\USBSTOR /v
   Start /t REG_DWORD /d 4 /f
```

10.2 Keyloggers & Hardware Spy Devices

Attackers may install hardware keyloggers to capture keystrokes, passwords, and sensitive data. These devices can be physically attached to keyboards, implanted in firmware, or wirelessly eavesdrop on keystrokes.

Common Keylogger Types:

- Inline USB Keyloggers – Small devices placed between a keyboard and a computer to record every keystroke.

- Wireless Keyloggers – Sniff keystrokes from wireless keyboards using radio interception (e.g., KeySweeper attacks).

- BIOS/UEFI Keyloggers – Malicious firmware implants that intercept keyboard input before the OS loads.

- Electromagnetic (TEMPEST) Keyloggers – Capture keystroke signals remotely using radio frequency analysis.

How to Defend Against Keyloggers:

- Physically inspect your keyboard cable for unusual inline devices.

- Use a tamper-evident seal on USB ports and keyboard connectors.

- Enable two-factor authentication (2FA) to prevent stolen passwords from being useful.

- Use an on-screen virtual keyboard for entering sensitive information.

- Use Faraday shielding to block electromagnetic (TEMPEST) surveillance of keystrokes.

Windows: Detecting Hidden Keyloggers

```
1  # List active keyboard devices (Windows PowerShell)
2  Get-PnpDevice | Where-Object { $_.Class -eq "Keyboard" }
```

10.3 Evil Maid Attacks & BIOS/UEFI Exploits

An Evil Maid Attack occurs when an attacker gains physical access to a system and installs malware or backdoors at the firmware level.

Common Evil Maid Attack Techniques:

- BIOS/UEFI Malware - Installs persistent malware at the firmware level.

- Cold Boot Attacks - Extracts encryption keys from RAM after a system is powered off.

- Bootloader Manipulation - Replaces the system bootloader with a keylogger or backdoor.

How to Defend Against Evil Maid Attacks:

- Enable Secure Boot to prevent unauthorized bootloader modifications.

- Use a BIOS/UEFI password to block unauthorized firmware changes.

- Encrypt the boot partition (LUKS2, BitLocker) to prevent bootloader tampering.

- Use a tamper-evident case (e.g., epoxy seals on screws and ports).

- Store laptops in a Faraday bag to prevent wireless wake-up attacks.

Advanced Defense

Consider using firmware security tools like Heads or NitroBoot to detect tampering at boot. These tools verify your system's integrity using cryptographic measurements before loading the operating system.

10.4 Malicious Peripherals: HID Spoofing & Firmware Attacks

Peripherals like keyboards, mice, and chargers can be weaponized with malicious firmware.

Example Threats:

- **BadUSB** - USB drives reprogrammed to act as keyboards.

- **OMG Cables** - Malicious charging cables with embedded wireless implants.

- **USB-C Chargers** - Hardware implants for malware delivery.

Countermeasures:

- Use **data-blocking adapters** (USB condoms) for safe charging.

- Purchase peripherals from **open-source or verified suppliers.**

- Disable **USB autorun** and enforce device whitelisting.

10.5 Side-Channel & Supply Chain Attacks

Side-channel attacks exploit leaked information from electronic devices, while supply chain attacks compromise hardware or firmware before it reaches the user.

Common Side-Channel & Supply Chain Attacks:

- TEMPEST Attacks — Monitors electromagnetic emissions to recover screen data or keystrokes.

- Spectre & Meltdown — Exploits CPU vulnerabilities to extract secret data from memory.

- Pre-Installed Backdoors — Firmware implants in motherboards, networking hardware, or security chips.

- Fake Charging Cables — Devices like O.MG Cable inject keyloggers or remote access payloads.

Defending Against Side-Channel & Supply Chain Attacks:

- Use Faraday shielding to prevent RF-based surveillance.

- Regularly update BIOS/UEFI firmware to patch backdoor exploits.

- Purchase hardware from trusted sources – Avoid cheap or unknown brands.

- Disable unused hardware features (Intel ME, AMD PSP, remote management tools).

- Use open-source hardware (Purism, System76, Nitrokey) where possible.

10.6 Final Hardware Security Recommendations

To protect against hardware-based attacks, follow these best practices:

- Physically secure devices – Use tamper-evident tape, case locks, and controlled access.

- Disable unused ports – Block USB, FireWire, and Thunderbolt ports in BIOS/UEFI.

- Monitor device integrity – Use hardware intrusion detection systems.

- Encrypt boot partitions – Prevent attackers from modifying system firmware.

- Regularly inspect devices – Look for unexpected modifications or implants.

By implementing these defenses, you can significantly reduce the risk of hardware-based attacks, whether from cybercriminals, malicious insiders, or nation-state actors.

11.

Secure Communication

In today's world, mass surveillance, metadata collection, and digital tracking make private communication difficult. Governments, corporations, and cybercriminals continuously monitor online activity, making end-to-end encryption (E2EE) essential for protecting personal and professional conversations.

Secure communication involves:

- End-to-End Encryption (E2EE) – Ensures only the sender and recipient can read messages.

- Metadata Protection – Prevents exposure of who, when, and where messages are sent.

- Decentralization – Reduces reliance on centralized services that can be subpoenaed or censored.

11.1 Threats to Private Communication

Even encrypted messages can still be vulnerable if metadata leaks, encryption keys are compromised, or backdoors exist in the platform.

Common Threats:

- Metadata Collection - Even encrypted apps like WhatsApp and Telegram store timestamps, contact lists, and IP addresses.

- Government Surveillance - Authorities subpoena messaging providers or use malware to bypass encryption.

- Compromised Encryption Keys - Weak key management can expose private conversations.

- Endpoint Attacks - Even secure chats can be intercepted if malware or keyloggers are installed on a device.

- Fake Encryption Apps - Attackers create backdoored "secure" chat apps (e.g., EncroChat, Anom) to trap users.

11.2 Choosing a Secure Messaging App

Not all encrypted messengers are equally private—some store metadata, while others are fully decentralized.

Recommended End-to-End Encrypted Messaging Apps:

- Signal - Best overall encrypted messenger; supports disappearing messages and phone number registration privacy.

- Session - Fully decentralized, onion-routed, and does not require a phone number.

- Briar - Works offline via Bluetooth or Wi-Fi mesh networks (ideal for high-risk environments).

- Element (Matrix) - Secure, decentralized chat with support for encrypted voice/video calls.

Avoid These Apps:

- WhatsApp - End-to-end encrypted but collects metadata and shares data with Meta (Facebook).

- Telegram - Encryption is not enabled by default, and metadata is retained.

- Skype / Zoom - Corporate surveillance and weak encryption history.

11.3 Using PGP for Secure Email Encryption

Most email services do not encrypt emails end-to-end, meaning providers can read, store, and analyze messages. Pretty Good Privacy (PGP) encryption ensures only the sender and recipient can access the email content.

How to Encrypt Emails Using GPG (GNU Privacy Guard):

```
1    # Generate a new PGP key
2    gpg --full-generate-key
3
4    # Export your public key to share with others
5    gpg --armor --export YOUR_EMAIL > mypublickey.asc
6
7    # Encrypt an email message
8    gpg --encrypt --recipient "RecipientKeyID" message.txt
9
10   # Decrypt a received email
11   gpg --decrypt message.txt.gpg > decrypted.txt
```

Best Practices for PGP:

- Use strong key sizes - RSA 4096-bit or ECC Curve25519.

- Regularly rotate encryption keys - Expire old keys and generate new ones periodically.

- Publish your public key carefully – Only share it with trusted contacts.

- Enable a key revocation certificate – Prevents abuse if your private key is lost or compromised.

11.4 Out-of-Band Key Verification

Encryption alone does not prevent man-in-the-middle attacks unless keys are properly verified. Out-of-band key verification ensures you are communicating with the intended person, not an attacker.

Best Verification Methods:

- **In-Person:** Meet physically to verify public key fingerprints.

- **Voice Call:** Read and verify key fingerprints aloud.

- **Multiple Channels:** Confirm keys over separate platforms (e.g., Signal and Email).

Warning: Never trust unverified keys, especially if the contact method is new or unexpected.

11.5 Self-Hosting Private Email Servers

For true privacy, self-hosting an email server is the best option. This prevents email providers from scanning messages or handing over data to governments.

Recommended Self-Hosted Email Solutions:

- Mail-in-a-Box – Open-source, simple-to-deploy personal email server.

- Poste.io – Secure, easy-to-use mail server with built-in spam filtering.

- Mailcow – Full-featured, Docker-based email suite.

If self-hosting isn't an option, consider privacy-respecting email providers:

Recommended Encrypted Email Services:

- ProtonMail – Swiss-based, zero-access encryption, supports PGP.

- Tutanota – End-to-end encrypted email with open-source security.

- CTemplar (if operational) – Anonymous email service with strong encryption.

11.6 Decentralized Social Media and Communication

Corporate-controlled platforms like Facebook, Twitter, and Instagram collect massive amounts of user data and censor content. To maintain privacy and freedom of speech, decentralized, censorship-resistant alternatives should be used.

Privacy-Focused Social Media Platforms:

- Mastodon – Decentralized alternative to Twitter/X, runs on the ActivityPub protocol.

- Nostr – Lightweight, censorship-resistant communication protocol.

- PeerTube – Open-source, federated alternative to YouTube.

Why Use Decentralized Platforms?

- No central authority can ban or censor accounts.

- Users retain full control over their data.

- Network resilience - No single point of failure.

11.7 Secure Multi-Party Communication Protocols

Collaborating with multiple trusted parties requires distributed trust.

Technologies:

- **Secret Sharing Schemes** (e.g., Shamir's Secret Sharing).

- **Secure Group Messengers** (e.g., Signal Communities, Matrix rooms).

Use Cases:

- Sharing sensitive keys without giving any one person full control.

- Coordinating multi-party disclosure or activism securely.

11.8 The Endpoint Problem: Device Compromise Risks

Even the strongest encryption fails if your device is compromised. Malware, keyloggers, and remote access trojans (RATs) can:

- Capture decrypted messages in real-time.

- Steal encryption keys or passphrases.

- Record screenshots or keystrokes.

Defensive Actions:

- Keep your operating system and apps updated.

- Use a reputable antivirus or anti-malware solution.

- Avoid installing unknown or untrusted software.

- Consider using a separate device for high-risk communications.

11.9 Traffic Analysis Resistance in Voice and Video Communication

Encrypted voice and video calls still leak timing and packet size metadata.

Threats:

- Timing analysis can reveal who is talking and when.

- Packet size patterns can infer speech activity.

Mitigation Techniques:

- Use **traffic padding** or **constant bitrate** applications.

- Prefer **p2p decentralized** messengers (Briar, Jami) with randomized packet timing.

11.10 Final Security Recommendations for Private Communication

To ensure maximum privacy when communicating online:

- Use end-to-end encrypted messaging apps (Signal, Session, Briar).

- Encrypt emails with PGP/GPG if discussing sensitive topics.

- Prefer decentralized and self-hosted solutions over Big Tech platforms.

- Use anonymous email services and avoid linking accounts.

- Enable disappearing messages for high-risk conversations.

- Regularly verify encryption keys to prevent man-in-the-middle attacks.

- Use Tor or a VPN when accessing secure communications.

By implementing these secure communication strategies, you can minimize your exposure to surveillance, data collection, and cyber threats.

12.

Physical Security & Anti-Forensics

Encryption is a beautiful thing—mathematically exquisite, unyielding to brute force, and designed to shield your data from prying eyes.

But let's not kid ourselves.

If someone gets their hands on your hardware—your laptop, your phone, your drive—you've already lost round one. Encryption? Still holding the line, yes. But physical access? That's when the real games begin. Cold boot attacks, firmware tampering, hardware implants…

You see, encryption is a vault. A good one. But even the best vaults sit in buildings that get robbed.

So while encryption buys you time—and it should always be your first line of defense—true security means stacking the deck in your favor. Physical controls. Tamper detection. Operational discipline. Because when your data's on the line, betting everything on a single safeguard isn't security. It's wishful thinking.

And you, dear reader, are far too clever to leave

it at that.

12.1 Physical Security Threats

If an attacker gains access to your devices, they can:

- Clone or extract data - Bypassing encryption through cold boot attacks or forensic tools.

- Install keyloggers - Hardware-based keyloggers or malware-infected USB devices.

- Tamper with firmware - Supply chain attacks, BIOS/TPM backdoors, or Evil Maid attacks.

- Track device location - Using IMSI catchers, GPS, or hidden Bluetooth trackers.

12.2 Decentralized Identity (DID) & Verifiable Credentials

Future identity systems may use cryptographically signed, user-controlled credentials.

Benefits:

- **No central authority** controls your identity.

- Supports **zero-knowledge proofs** for privacy-preserving verification.

Example Projects:

- W3C Decentralized Identifier (DID) Standard.

- Verifiable Credential protocols for selective disclosure.

12.3 Securing Your Devices Against Physical Attacks

To prevent tampering, surveillance, or theft, apply the following measures:

1. Use a Faraday Bag for Laptops and Phones

- Blocks GPS, Bluetooth, Wi-Fi, RFID, and mobile network signals.

- Prevents remote hacking and tracking while traveling.

- Best options: **Mission Darkness, Silent Pocket, SLNT.**

2. Enable Full-Disk Encryption (FDE)

- Protects data from unauthorized access even if the device is stolen.

- Recommended tools:

  - Linux - LUKS/dm-crypt
  - Windows - VeraCrypt (BitLocker is not fully trusted)
  - MacOS - FileVault

3. Apply Tamper-Evident Seals

- Use security tape or forensic evidence seals to detect unauthorized access.

- Place seals over laptop screws, USB ports, and battery compartments.

- Change seals regularly and record serial numbers for verification.

4. Use Anti-Tamper Screws and Security Hardware

• Prevent unauthorized modifications with Torx, security, or pentalobe screws.

• Apply physical locks (Kensington lock, padlocks) to laptops and servers.

5. Defend Against "Evil Maid" Attacks

• Evil Maid attacks involve installing malware on unattended devices.

• To prevent:

  - Use a BIOS password – Prevent unauthorized firmware changes.

  - Enable TPM + Secure Boot – Prevent untrusted OS tampering.

  - Use Qubes Anti-Evil Maid (AEM) – Verifies boot integrity.

  - Store your laptop in a safe or lockbox when unattended.

6. Detect and Prevent Hidden Trackers

• IMSI Catchers & Stingrays – Used by law enforcement to intercept mobile signals.

  - **Countermeasures:** Use "SnoopSnitch" (Android) or a Faraday bag.

• AirTags, Tile, and GPS Trackers – Used for covert surveillance.

  - **Countermeasures:** Scan for Bluetooth anomalies with "LightBlue" or "Find My".

7. Protect Against Surveillance Cameras and Spy Devices

- Hidden cameras and audio recorders can compromise privacy.

- **Detection tools:**

  - Use an RF Detector (e.g., JMDHKK) to scan for active signals.
  - Use an IR camera to detect hidden night vision cameras.
  - Turn off voice assistants (Alexa, Google Assistant, Siri).

12.4 Detecting Network Taps and Implants

Surveillance implants can exist in network cables, routers, and infrastructure.

Detection Techniques:

- Visually inspect for **tap splitters or fiber bends.**

- Use **network monitoring tools** to detect mirrored traffic.

- Scan for **RF emissions** using spectrum analyzers.

Devices can be intercepted or tampered with before they even reach you. Known as supply chain interdiction, this risk is particularly high for:

Activists

Journalists

Corporate or government targets

High-value individuals or organizations

Example Threats:

• Pre-installed keyloggers or implants.

• Backdoored firmware or BIOS.

• Hardware modified to include surveillance chips.

Mitigation Strategies:

• **Source Hardware from Trusted Vendors:** Prefer vendors with strict supply chain security protocols.

• **Verify Firmware Signatures:** Re-flash BIOS with known good firmware where possible.

• **Use Hardware Provenance Auditing:** For high-risk deployments, log serial numbers and shipping paths to detect anomalies.

• **Rebuild From Source:** If you have the expertise, recompile firmware (Libreboot, Coreboot) to remove backdoors.

Even full-disk encryption can be defeated if an attacker extracts keys from RAM during or shortly after shutdown.

Advanced Defenses:

• **Power Down Completely:** Fully shut down the system before leaving it unattended. Suspend or hibernate modes leave keys in RAM.

• **Disable Sleep-to-Disk (Hibernate):** On Linux, set nohibernate or disable swap.

- **Memory Zeroing on Shutdown:** Use OS-level tools or kernel modules that scrub RAM during shutdown (e.g., kernel.core_pattern, memwipe, or secure shutdown hooks).

- **Use RAM Encryption Features:** On supported hardware (e.g., AMD Secure Memory Encryption, Intel TME), enable hardware-based RAM encryption.

Best Practices When Crossing Borders or Traveling in Hostile Areas:

- **Use a Clean Burner Device:** Carry a wiped, encrypted laptop or phone with no personal data.

- **Remove All Non-Essential Data:** Backup and erase anything you don't want to lose or reveal.

- **Prepare Travel-Specific Accounts:** New email, messaging, and social accounts tied to burner identities.

- **Encrypt and Power Down:** Always power off and encrypt devices before customs, border crossings, or checkpoints.

- **Use Data Destruction Safeguards:** Set up self-destruct mechanisms or duress passwords on sensitive devices.

12.5 Secure Data Destruction

Simply deleting files is not enough. To prevent forensic recovery, data must be securely erased or physically destroyed.

1. Wiping HDDs and SSDs Securely

For HDDs (Magnetic Disks):

```
1  # Securely wipe an entire hard drive (overwrite with random
   data)
2  shred -v -n 5 -z /dev/sdX
```

For SSDs (Flash Storage):

- Use "blkdiscard" to wipe flash memory:

```
1      sudo blkdiscard /dev/sdX
```

- Enable TRIM and secure erase using the SSD manufacturer's tool.

2. Physically Destroying Storage Devices

- Degaussing - Use an industrial-grade degausser to destroy HDDs.

- Drilling & Shredding - Physically drill through the platters or use an industrial shredder.

- Thermite Destruction - Pouring thermite on a drive will burn through metal and silicon.

- Microwave (for flash storage) - Destroy USB drives and SD cards in a high-powered microwave.

- EMP (Electromagnetic Pulse) - Can render SSDs and flash memory unreadable.

12.6 Forensic Anti-Analysis Techniques

Even if a device is seized, forensic experts may attempt to recover data. Use the following anti-forensics techniques to minimize exposure.

1. Use "Deniable Encryption" to Conceal Data

- VeraCrypt Hidden Volumes - Allows plausible deniability by creating a fake partition.

- LUKS2 with Detached Header - Separates encryption metadata from the drive.

2. Prevent Cold Boot Attacks

- Cold boot attacks extract encryption keys from RAM after a sudden power loss.

- To prevent:

  - **Power off the system completely** before traveling.
  - Use anti-forensics memory scrubbing tools (e.g., TRESOR).

3. Disable Forensic Data Collection on Phones

- **Factory Reset is NOT enough** - Flash a new OS after a reset.

- Use **GrapheneOS or CalyxOS** - Prevents forensic extraction tools like Cellebrite.

- Disable USB Debugging to prevent forensic mode access.

12.7 Physical Security Recommendations

To maintain full physical security and prevent forensic recovery:

- Encrypt all devices with strong passphrases.

- Use tamper-evident seals to detect unauthorized access.

- Physically destroy old drives before disposal.

- Travel with burner devices instead of your primary system.

- Assume all surveillance cameras and microphones are recording.

By following these security practices, you can ensure your data remains safe, even in worst-case scenarios.

13.

Anti-Forensics & Data Destruction

Some people secure data to avoid being profiled by advertisers. Others secure data because losing it could change the course of history.

If you're reading this chapter, you likely fall into the second category.

13.1 Why Advanced Anti-Forensics Matters

Digital forensics has evolved into an industry of its own. State-sponsored actors, private contractors, and law enforcement agencies have access to tools like:

- **Cellebrite** – Extracts data from encrypted smartphones.

- **Magnet AXIOM** – Reconstructs digital timelines from logs, metadata, and deleted files.

- **XRY, Oxygen Forensics, and GrayKey** – Bypass device locks and encryption on mobile and desktop systems.

These tools don't just recover data—they reconstruct your behavior, contacts, and intent.

13.2 Secure Deletion: Erasing vs. Destroying

Secure deletion is the first layer of defense, but remember:

> *Data you delete may not actually be gone—unless you destroy it.*

Overwriting HDDs vs. SSDs

- **HDDs (Magnetic Drives):** Multiple overwrites (e.g., using shred) can effectively sanitize magnetic disks.

- **SSDs (Flash Storage):** Due to wear-leveling, overwriting isn't reliable. Always use full-disk encryption from day one, then destroy the drive physically when done.

Secure Deletion Methods: Software

HDDs vs. SSDs: Know Your Medium HDDs (Spinning Disks)

Overwrite with random data at least 3-5 passes.

- Example: shred -v -n 5 -z /dev/sdX

SSDs (Solid State)

- Overwriting doesn't reliably hit all sectors.

- Use blkdiscard or the manufacturer's secure-erase tools.

- Full-disk encryption with ephemeral keys is best—wipe the keys, the data dies with them.

Physical Destruction Methods

- **Shred and Pulverize:** Industrial shredders or hammer and drill attacks.

- **Degaussing:** Industrial electromagnets to destroy magnetic storage.

- **Thermite or Incineration:** When plausible deniability isn't an option, total destruction is the only guarantee.

- High-powered microwaves (for USB sticks or SIMs—*safely, please*).

- Portable EMP generators (for the truly committed).

13.3 Forensic Evasion Tradecraft

Digital forensics is about timelines, metadata, and artifacts. **Anti-forensics is about breaking those timelines.**

Hidden Encrypted Volumes

VeraCrypt's hidden volumes provide plausible deniability by:

- Allowing two passwords: one for decoy data, one for real data.

- Providing no cryptographic evidence that hidden data exists.

LUKS2 with Detached Header: Encryption header stored separately—looks like random junk without the header file.

    Operational Tip: Populate the decoy volume with realistic, non-critical data to sell the illusion.

Memory Scrubbing Tools

Tools like **TRESOR** move encryption keys to CPU registers, preventing cold boot attacks that recover keys from RAM. **Operational Tip:** Always perform a full shutdown—not sleep or hibernate—when moving across hostile borders.

Timestamp and Metadata Obfuscation

Forensic timelines depend on metadata. You can:

- Use `touch` or `setfattr` to manipulate file timestamps.

- Clone files with altered dates to muddy the timeline.

- Use `mat2` or `exiftool` to strip metadata from documents, images, and media.

13.4 Deniable Encryption & Steganography

When destruction isn't possible, deception becomes your strongest ally.

Fake Decoy Data

Populate systems with believable—but harmless—data that appears valuable. This misdirection can buy time or satisfy adversaries during coercion.

Steganography: Hiding in Plain Sight

Steganography allows you to embed data inside:

- Images (JPEG, PNG)
- Audio files (MP3, WAV)
- Video or document files

Example using steghide:

```
1   # Hide secret.txt inside cover.jpg
2   steghide embed -cf cover.jpg -ef secret.txt
```

Retrieving hidden data:

```
1   steghide extract -sf cover.jpg
```

13.5 Air-Gapped Storage and Dead Drops

Air-gapped backups stored offline are nearly impossible to extract without physical access. Consider using:

- Encrypted microSD cards stored in tamper-evident containers.
- Hidden physical dead drops for multi-party data exchange.

Operational Tip: Use Faraday shielding to prevent wireless data exfiltration from hidden devices.

13.6 Advanced Data Safeguards the Feds Hate

1. Air-Gapped Backups No network connectivity. Stored offline, encrypted. Ideal for whistleblowers or data you can't afford to lose—or have found.

2. One-Time Cryptographic Erasure Ephemeral keys with no backup. Delete the key, destroy the data. No second chances.

3. Live OS with No Persistence (Tails or Qubes Disposable VMs) Boot from USB, leave no trace. Pull the stick, data's gone. Useful for: Anonymous research. One-way communication drops. High-risk fieldwork.

13.7 Final Operational Reality Check

Anti-forensics isn't about making data disappear. It's about *controlling what your adversary sees*—or doesn't see—when they inevitably come looking.

The harsh reality? **If you can't control the narrative, control the evidence.**

14.

Hardened Operating Systems (Linux)

The operating system is the foundation of digital security. Using a privacy-focused, hardened OS significantly reduces vulnerabilities.

14.1 Best Linux Distributions for Security

Depending on your threat model, choose a security-oriented OS:

- Qubes OS - **Best for ultra-secure users.** Uses a compartmentalized (Xen-based) system where applications run in isolated VMs.

- Tails OS - **Best for anonymity.** Live OS that routes all traffic through Tor, leaving no trace.

- Linux Kodachi - **Best for privacy.** A powerful privacy and anonymity operating system. One my favorite and one of the most underrated Linux distros.

- Parrot OS - **Balanced between privacy and pentesting.** Includes sandboxing, Tor, and forensic tools.

- Whonix - **Best for Tor anonymity.** Uses two VMs (a Tor gateway and a workstation) to isolate traffic.

- GrapheneOS (Mobile) - **Most secure Android alternative.** De-Googled OS with advanced security patches.

14.2 Application Sandboxing: Firejail, Bubblewrap, Flatpak

Applications can leak sensitive data or be exploited. Use sandboxing to isolate them.

Using Firejail (Lightweight Sandboxing):

```
1   # Run Firefox in a sandbox
2   firejail firefox
3   # Check restrictions
4   firejail --list
```

Using Flatpak (Secure App Containment):

```
1   # Install Flatpak
2   sudo apt install flatpak
3
4   # Run an app in a sandbox
5   flatpak run org.mozilla.firefox
```

Bubblewrap (Micro-Isolation for Linux Apps):

```
1   bwrap --ro-bind / / --dev /dev --proc /proc --chdir /home/
      user firefox
```

14.3 Hardening Linux for Security and Privacy

Disable Unnecessary Services:

```
1  sudo systemctl disable bluetooth
2  sudo systemctl mask bluetooth.service
3  sudo systemctl disable ssh
```

Seccomp (System Call Filtering):

```
1  # Firejail uses seccomp to block dangerous system calls by
   default
2  firejail --seccomp firefox
```

Harden the Kernel and System Configurations:

```
1  # Prevent kernel exploits
2  echo "kernel.kptr_restrict = 2" | sudo tee -a /etc/sysctl.
   conf
3  # Prevent buffer overflow attacks
4  echo "kernel.randomize_va_space = 2" | sudo tee -a /etc/
   sysctl.conf
5
6  # Apply changes
7  sudo sysctl -p
```

Enable AppArmor for Application Security:

```
1  # Check AppArmor status
2  sudo aa-status
3
4  # Enforce security policies
5  sudo aa-enforce /etc/apparmor.d/\\
```

Enable System Auditing (Tracks Unauthorized Access):

```
1  # Install auditd (Linux Audit System)
2  sudo apt install auditd
3
4  # Monitor all authentication attempts
5  sudo auditctl -w /var/log/auth.log -p wa -k auth_attempts
```

> **Note**
>
> While Flatpak provides sandboxing, always review the requested permissions. Not all Flatpak apps are fully isolated, and some remotes may distribute poorly configured packages.

Block All Incoming Network Traffic:

```
# Block all inbound traffic with iptables
sudo iptables -P INPUT DROP
sudo iptables -P FORWARD DROP
sudo iptables -A INPUT -m conntrack --ctstate ESTABLISHED,
    RELATED -j ACCEPT
```

14.4 Hardening Linux: From Hobbyist to Hardened Operator

Disable What You Don't Need (Because Attackers Will Use It) Bluetooth? Off.

SSH? Off if you aren't using it.

Avahi, CUPS, Samba? Kill anything you don't control.

Kernel and Memory Protections Enable kernel pointer restriction.

Enforce memory randomization (ASLR).

Apply sysctl lockdowns.

Leverage Mandatory Access Control (AppArmor/SELinux) AppArmor: Easier to manage, great for Ubuntu/Debian-based systems.

SELinux: More granular, enterprise-grade, but complex.

If it's not enforcing, it's not protecting.

System Auditing: Know Who Touched What, and When
Monitor logins.

Track file changes.

Alert on suspicious access.

14.5 Full Disk Encryption with LUKS

Full disk encryption protects your data at rest,
especially if your device is lost or stolen.

Setting Up LUKS Encryption on a New Partition:

```
1   # Install cryptsetup if not already installed
2   sudo apt install cryptsetup
3
4   # Initialize LUKS on a partition (WARNING: This will destroy
        data on /dev/sdX)
5   sudo cryptsetup luksFormat /dev/sdX
6
7   # Open the encrypted partition
8   sudo cryptsetup luksOpen /dev/sdX secure_partition
9
10  # Create a filesystem
11  sudo mkfs.ext4 /dev/mapper/secure_partition
12
13  # Mount it
14  sudo mount /dev/mapper/secure_partition /mnt
```

> **Warning**
>
> LUKS setup will erase all data on the selected
> partition. Backup before proceeding.

Most Linux distributions support full disk encryption
during installation. Always enable it on laptops or
portable systems.

14.6 Why This Matters More Than Ever

Governments build exploits. Corporations build telemetry. Attackers build botnets.

You? You build resilience.

Your OS isn't just the system you run— It's the territory you defend. Choose wisely. Harden relentlessly. Because once they're in, there is no patch for regret.

15.

Windows Hardening

While this manual primarily focuses on privacy and security within open-source and Linux-based environments, many users still rely on Windows and iOS in their daily lives. These operating systems are heavily monitored and come with built-in telemetry, tracking, and security vulnerabilities. This section provides practical steps to lock down Windows and iOS for better privacy and security.

15.1 Windows Hardening

Let's be clear—Windows doesn't respect your privacy out of the box. You paid for the license, and Microsoft still treats you like the product. But don't worry, you're not helpless. You just have to be willing to say "No" more aggressively than Windows wants you to.

15.2 Disable Microsoft Telemetry and Tracking

Microsoft collects vast amounts of diagnostic and usage data by default. Disable unnecessary telemetry by using the following steps:

- Use O&O ShutUp10++ - A free tool that disables Windows spying and telemetry. (https://www.oo-software.com/en/shutup10)

- Or Simplewall—the firewall Microsoft should have built but didn't, because telemetry has shareholders to impress.

- To manually disable telemetry:

  1. Open Settings → Privacy → Diagnostics and Feedback.

  2. Set "Send diagnostic data" to Basic.

  3. Disable "Improve inking and typing recognition".

  4. Turn off "Tailored experiences" and "Advertising ID tracking".

Windows comes with a significant amount of pre-installed bloatware, telemetry, and unnecessary services that degrade performance and compromise privacy. Debloating your system can improve speed, security, and overall usability. However, always be cautious when running scripts that modify system settings.

The following Windows debloating scripts have been tested and verified on personal and work-related devices:

- Simeonon Security Windows Debloat Script https://github.com/simeononsecurity/Windows-Opt imize-Debloat A well-maintained script for removing unnecessary Windows components while preserving core functionality.

- Simeonon Security Windows Debloat and Hardening Script https://github.com/simeononsecurity/Wind ows-Optimize-Harden-Debloat Combines debloating with security enhancements for better system protection.

- Chris Titus' Perfect Install Script Run the following command in an Administrator PowerShell Terminal: "'powershell irm christitus.com/win | iex "' This script automates Windows optimization, privacy enhancements, and software installations.

- Raphire Windows 11 Debloat Script https://gith ub.com/Raphire/Win11Debloat A lightweight script designed specifically to debloat and optimize Windows 11.

15.3 Firewall and Network Hardening

The default Windows firewall is not sufficiently restrictive. Instead:

- Use Simplewall to block all unwanted network activity (https://github.com/henrypp/simplewall).

- Disable Remote Assistance, Remote Desktop, and SMBv1 to prevent remote exploits.

- Use a DNS-over-HTTPS (DoH) provider like NextDNS or Quad9 for encrypted DNS queries.

15.4 Application and Software Security

Avoid installing unnecessary applications. Follow these hardening steps:

- Uninstall bloatware: Run 'Get-AppxPackage | Remove-AppxPackage' in PowerShell to remove unnecessary default apps.

- Use a Limited User Account instead of an Administrator account for daily use.

- Disable Windows Script Host to prevent script-based malware attacks.

- Use VeraCrypt to encrypt your hard drive instead of relying on BitLocker.

15.5 Secure Web Browsing and Communication

Use privacy-focused browsers and messaging apps:

- Browser Hardening:

  - Use LibreWolf or Brave instead of Edge or Chrome.

  - Enable uBlock Origin and LocalCDN extensions.

  - Use private search engines like DuckDuckGo or Startpage.

- Encrypted Messaging:

  - Use Signal for private messaging and calls.

  - Avoid SMS-based 2FA; instead, use TOTP (Aegis, Authy).

15.6 Enable Exploit Protection

Windows includes system-wide exploit mitigations
like Data Execution Prevention (DEP) and Address
Space Layout Randomization (ASLR).

- Open Windows Security → App & Browser Control
  → Exploit Protection.

- Enable system-wide ASLR and Control Flow Guard
  (CFG).

15.7 Enable Credential Guard & LSA Protection

Credential Guard helps protect against credential
dumping tools like **Mimikatz** by isolating secrets in
a protected memory environment. LSA (Local Security
Authority) Protection prevents malicious processes
from accessing credentials stored in memory.

How to Enable:

- Go to **Windows Security** → **Device Security** → **Core
  Isolation**.

- Enable **Memory Integrity**.

Or enable LSA Protection via Registry:

```
1
2   Enable LSA Protection (RunAsPPL)
3   reg add "HKLM\SYSTEM\CurrentControlSet\Control\Lsa" /v "
        RunAsPPL" /t REG_DWORD /d 1 /f
```

15.8 Disable Legacy SMBv1 Protocol

SMBv1 is a severely outdated protocol targeted
by attacks like **EternalBlue**. Disable it unless

absolutely required for legacy compatibility.

```
1
2  Disable SMBv1 Protocol
3  Disable-WindowsOptionalFeature -Online -FeatureName "
      SMB1Protocol"
```

15.9 Block LSASS Memory Credential Dumping

Prevent attackers from extracting plaintext credentials by hardening WDigest settings.

```
1
2  Disable WDigest Credential Caching
3  reg add "HKLM\SYSTEM\CurrentControlSet\Control\
      SecurityProviders\WDigest" /v "UseLogonCredential" /t
      REG_DWORD /d 0 /f
```

This prevents credential material from being left accessible in memory.

15.10 Lock Down App Execution with WDAC

Advanced users can enable Windows Defender Application Control (WDAC) to whitelist trusted applications and block all others by default.

- Create a Code Integrity Policy using Microsoft's tools.

- Sign and enforce allowed applications only.

- Prevent unsigned or unknown software from running.

Caution: WDAC can lock you out if misconfigured. Test in a virtual machine first.

15.11 Tighten SmartScreen or Disable Script Hosts

Microsoft SmartScreen checks downloads and apps against Microsoft's reputation database. You can tighten or disable it based on your trust model.

- Go to Windows Security → App & Browser Control.

- Set SmartScreen to "Warn" or "Block" for apps and downloads.

Disable Windows Script Host to block malicious scripts:

```
1   # Disable Windows Script Host via Registry
2   reg add "HKLM\Software\Microsoft\Windows Script Host\
        Settings" /v Enabled /t REG_DWORD /d 0 /f
```

15.12 Harden Microsoft Defender

While not perfect, Microsoft Defender can be hardened by enabling Cloud-Delivered Protection, enabling Automatic Sample Submission and by turning on Tamper Protection to prevent malware from disabling Defender.

16.

iOS Hardening

iOS is generally more locked-down than Android, but it still tracks users extensively through iCloud, Siri, and default settings. Here's how to harden an iPhone:

16.1 Disable Apple Telemetry and Tracking

- Go to **Settings** → **Privacy & Security** → **Analytics & Improvements**.

- Disable all options under **Analytics & Improvements**.

- Turn off **Personalized Ads** in **Apple Advertising**.

- Disable **Significant Locations** (used for location tracking).

- Go to **Settings** → **Siri & Search** and disable **"Listen for 'Hey Siri'"** to prevent passive microphone activation.

16.2 Hardening Safari and Web Browsing

Safari has built-in tracking protection, but it needs further hardening:

- Enable **Prevent Cross-Site Tracking** in Safari settings.

- Disable **Motion & Orientation Access** to prevent device fingerprinting.

- Set search engine to **DuckDuckGo** or **Startpage**.

- Block all third-party cookies and disable website tracking.

- Use **Brave** or **Mullvad Browser** instead of Safari for sensitive browsing.

16.3 Locking Down iCloud and Apple Services

iCloud is convenient. It's also a law enforcement backdoor disguised as a feature. iCloud automatically stores a vast amount of personal data. If you must use iCloud:

- Disable **iCloud Keychain** and use an offline password manager like **KeePassXC**.

- Turn off **Find My iPhone** if privacy is a priority.

- Use **Mail Privacy Protection** to prevent tracking pixels in emails.

- Avoid iCloud backups, Apple can decrypt them on request.

- Use **cryptographic cloud storage** like Proton Drive, Tresorit, or self-hosted Nextcloud.

16.4 iPhone Security and Network Protection

- Enable **Lockdown Mode** (iOS 16+) for high-risk scenarios.

- Disable **Wi-Fi Auto-Join** to prevent rogue network attacks.

- Use **Mullvad VPN** or **ivPN** for private internet access.

- Block unknown **AirDrop requests** to prevent AirDrop-based exploits.

- Use **USB Restricted Mode** (Settings → Face ID & Passcode) to prevent USB-based attacks.

- Keep the **camera and microphone blocked** when not in use.

16.5 Jailbreaking: Security Risks vs. Privacy Benefits

Jailbreaking provides deeper control over iOS, but it also weakens system integrity, making devices vulnerable to malware.

16.6 Risks of Jailbreaking

- Removes Apple's built-in security layers, increasing malware risk.

- Jailbreak tweaks may contain hidden vulnerabilities.

- Exposes SSH remotely if configured improperly.

16.7 Safe Jailbreak Hardening (For Advanced Users)

If jailbreaking is necessary, use only **trusted security-focused tweaks**:

- **LittleSnitch Mobile:** Prevents apps from secretly connecting to trackers.

- **Liberty Lite:** Blocks jailbreak detection in apps.

- **AppFirewall:** Provides fine-grained network control.

- **Tor-enabled browsers** for enhanced anonymity.

16.8 Detecting AirTags and Bluetooth Trackers

Stalkerware isn't limited to software. Apple's AirTags and other Bluetooth tracking devices can silently follow you.

16.9 How AirTag Stalking Works

- AirTags broadcast Bluetooth signals that nearby Apple devices relay to iCloud.

- Someone can plant an AirTag on you or your vehicle without your knowledge.

16.10 How to Detect AirTags and Bluetooth Trackers

- iOS automatically alerts users when an unknown AirTag is moving with them.

- Use **Tracker Detect** on Android to scan for nearby Apple trackers.

- Physically inspect bags, clothing, and vehicles for planted devices.

For other Bluetooth trackers like Tile, use Bluetooth scanning apps such as:

- **LightBlue** (iOS/Android)

- **Bluetooth Scanner** (Android)

16.11 Covert Privacy Techniques

For those in high-risk environments (activists, journalists, whistleblowers), additional precautions are required:

- Use an **Anonymous Apple ID** created over Tor.

- **Disable Face ID / Touch ID** in high-risk areas (Power + Volume Up).

- Keep the phone in a **Faraday bag** when not in use.

- Use a secondary, clean **burner phone** for sensitive activities.

- Avoid using the device for both personal and high-risk activities.

iOS and Windows were never designed for privacy, but with these changes, they can be hardened against surveillance and exploitation.

Final Reality Check

Apple sells privacy as a feature—but it's privacy *on their terms*.

You can disable telemetry, harden settings, and avoid iCloud, but the walled garden still has a gatekeeper.

You're not locking down an open platform. You're plugging leaks in a ship that was designed to float you into their ecosystem.

Make it work for you. But never mistake it for your territory.

17.

Automated Hardening Scripts

Manually applying security policies is time-consuming. Use automation scripts to enhance system security efficiently.

17.1 Linux Security Automation Script

```
1   #!/bin/bash
2   echo "Starting Linux Hardening Script..."
3   echo "Disabling Bluetooth..."
4   sudo systemctl disable bluetooth
5   # Enforce AppArmor
6   echo "Enforcing AppArmor..."
7   sudo systemctl enable apparmor
8   echo "Enabling security updates..."
9   sudo apt install unattended-upgrades -y
10  sudo dpkg-reconfigure --priority=low unattended-upgrades
11  # Set up a strict firewall policy
12  echo "Configuring UFW Firewall..."
13  sudo ufw default deny incoming
14  sudo ufw default allow outgoing
15  sudo ufw enable
16  echo "Linux hardening completed."
```

Listing 17.1: Linux Hardening Script

The above script applies **basic hardening measures** like disabling weak services, setting up a firewall, and enabling security controls.

17.2 Advanced SSH Security Script

SSH is a common attack vector. Harden it by disabling password authentication, enforcing key-based authentication, and using Fail2Ban to prevent brute-force attacks.

```
1  #!/bin/bash
2  echo "Securing SSH..."
3  # Disable root login over SSH
4  sudo sed -i 's/#PermitRootLogin yes/PermitRootLogin no/' /
     etc/ssh/sshd_config
5  # Disable password authentication (force key-based
     authentication)
6  sudo sed -i 's/#PasswordAuthentication yes/
     PasswordAuthentication no/' /etc/ssh/sshd_config
```

Listing 17.2: SSH Security Script

```
1  # Restart SSH service to apply changes
2  sudo systemctl restart sshd
3
4  # Install Fail2Ban to prevent brute-force attacks
5  sudo apt install fail2ban -y
6
7  echo "SSH hardening completed."
```

17.3 Network Hardening Script

This script blocks unnecessary ports, enables secure DNS (DNS-over-HTTPS), and prevents network scanning attacks.

```
1   #!/bin/bash
2   echo "Starting Network Hardening..."
3
4   # Block all unused ports except SSH, HTTPS, and DNS
5   sudo ufw default deny incoming
6   sudo ufw allow 22/tcp    # Allow SSH
7   sudo ufw allow 443/tcp   # Allow HTTPS
8   sudo ufw allow 53/udp    # Allow DNS
9
10  # Enable DNS-over-HTTPS (DoH) with Cloudflare
11  echo "server=1.1.1.1#cloudflare-dns.com" | sudo tee -a /etc/
        dnsmasq.conf
12  sudo systemctl restart dnsmasq
13
14  echo "Network hardening completed."
```

Listing 17.3: Network Hardening Script

17.4 Final Hardening Recommendations

After running these scripts, apply additional security measures:

- Use strong passwords and 2FA for all accounts.

- Disable unnecessary kernel modules (USB, Bluetooth, Webcam).

- Enable encrypted backups with BorgBackup, Restic, or VeraCrypt.

- Monitor system logs regularly for unusual activity:

```
1   sudo journalctl -p 3 -xb
```

By using these hardened operating systems and automated security scripts, you significantly reduce attack surfaces and improve digital security.

18.

Self-Hosting & Private Cloud Storage

Why self-host? Cloud services like Google Drive and Dropbox scan your data and log metadata. Self-hosting your own cloud storage gives you **complete control over your files** while improving privacy and security.

18.1 Best Private Cloud Storage Options

- Nextcloud – Open-source alternative to Google Drive with file sharing, calendars, and notes.

- Seafile – Performance-optimized, privacy-focused file sync server.

- ownCloud – Similar to Nextcloud but optimized for enterprise use.

- Tahoe-LAFS – A distributed cloud storage system with built-in encryption.

- IPFS (InterPlanetary File System) – Decentralized, censorship-resistant file storage.

18.2 Choosing the Best Cloud Alternative

Cloud Alternative	Best For	Security Features
Nextcloud	General self-hosted cloud	End-to-end encryption, file versioning, user management
Seafile	High-performance file syncing	Server-side encryption, faster sync speeds
ownCloud	Enterprise-grade cloud storage	LDAP/AD support, scalable architecture
Tahoe-LAFS	Decentralized, distributed storage	Fully encrypted, fault-tolerant network
IPFS	Peer-to-peer cloud storage	Censorship-resistant, globally distributed network

Table 18.1: Comparison of Self-Hosted Cloud Storage Solutions

18.3 Advanced Self-Hosting Security Measures

1. Use a Reverse Proxy for SSL Encryption

```
# Install Nginx
sudo apt install nginx

# Configure an HTTPS reverse proxy for Nextcloud
sudo nano /etc/nginx/sites-available/nextcloud

server {
  listen 443 ssl;
  server_name yourdomain.com;
  ssl_certificate /etc/letsencrypt/live/yourdomain.com/
      fullchain.pem;
  ssl_certificate_key /etc/letsencrypt/live/yourdomain.com/
      privkey.pem;
  location / {
    proxy_pass http://127.0.0.1:8080;
    proxy_set_header X-Forwarded-For $remote_addr;
  }
}
```

2. Enable Encrypted Backups for Your Cloud Data

```
sudo apt install borgbackup
borg init --encryption=repokey /mnt/backup
borg create --progress /mnt/backup::nextcloud-backup /var/
    www/html
```

3. Secure Nextcloud with Fail2Ban Rules

```
# Create a fail2ban filter for Nextcloud brute-force
    protection
sudo nano /etc/fail2ban/filter.d/nextcloud.conf

[Definition]
failregex = .*"message":"Login failed: .* Remote IP: <HOST>"
```

18.4 Setting Up a Private Nextcloud Server

Nextcloud is the best open-source alternative to Google Drive.

Installation on Ubuntu (Basic Setup):

```
1  sudo apt update
2  sudo apt install nextcloud apache2 mariadb-server php php-
     mysql
3  sudo systemctl enable apache2 mariadb
4  sudo systemctl start apache2 mariadb
5  sudo mysql_secure_installation
```

Alternative: Nextcloud with Docker (More Secure):

```
1  # Pull and run Nextcloud with PostgreSQL and Redis for
     better performance
2  docker run -d -p 8080:80 --name nextcloud \
3  -e POSTGRES_HOST=db -e POSTGRES_DB=nextcloud \
4  -e POSTGRES_USER=nextcloud -e POSTGRES_PASSWORD=securepass \
5  -v /opt/nextcloud:/var/www/html \
6  nextcloud
```

Securing Nextcloud:

- Use Let's Encrypt TLS Certificates for HTTPS encryption.

```
1     sudo apt install certbot python3-certbot-apache
2     sudo certbot --apache -d yourdomain.com
```

- Enable fail2ban to block brute-force attacks.

```
1     sudo apt install fail2ban
2     sudo systemctl enable fail2ban
```

- Harden database security with PostgreSQL over MariaDB.

18.5 Encrypted File Syncing with Syncthing

Syncthing is a peer-to-peer encrypted file syncing solution that allows users to securely synchronize files across devices without a centralized server.

Installing Syncthing on Linux:

```
1   # Install Syncthing
2   sudo apt install syncthing
3
4   # Start Syncthing service
5   syncthing --no-browser
```

Configuring Syncthing for Security:

- Disable relay servers to prevent third-party metadata logging.

```
1       sed -i 's|"relaysEnabled": true|"relaysEnabled":
        false|' ~/.config/syncthing/config.xml
```

- Restrict web interface to localhost (prevents remote access).

```
1       sed -i 's|<address>0.0.0.0:8384</address>|<address
        >127.0.0.1:8384</address>|' ~/.config/syncthing/
        config.xml
```

- Enforce TLS encryption for all file transfers.

18.6 Self-Hosting a Password Manager (Vaultwarden)

Vaultwarden is a lightweight, self-hosted alternative to Bitwarden.

Install Vaultwarden with Docker:

```
1   docker run -d --name vaultwarden -v /vw-data:/data -p
        8080:80 vaultwarden/server
```

Security Tips:

- Enable HTTPS with a reverse proxy (Nginx + Certbot).

- Require strong master passwords and 2FA.

18.7 Final Recommendations

Best Self-Hosting Practices:

- Use a dedicated server (Raspberry Pi, VPS, or local machine).

- Always enable full-disk encryption for self-hosted storage, and secure with firewalls, fail2ban, and intrusion detection.

- Encrypt backups using BorgBackup, Restic, or Duplicati.

- Regularly apply security updates to self-hosted services.

By self-hosting your cloud storage, you eliminate third-party surveillance and retain full control over your data.

19.

Self-Hosting Privacy Tools

Why self-host? Centralized cloud providers (Google Drive, Dropbox, Gmail) log, track, and monetize your data. Self-hosting ensures full control, better security, and stronger privacy.

19.1 Private Cloud Storage with Nextcloud

Nextcloud is an open-source cloud storage solution that allows users to store and sync files securely.

Installing Nextcloud on Ubuntu:

```
1  # Update system
2  sudo apt update && sudo apt upgrade -y
3  # Install Nextcloud and dependencies
4  sudo apt install nextcloud mariadb-server php php-mysql
5
6  # Enable Nextcloud in Apache
7  sudo systemctl enable apache2
```

Alternative: Secure Nextcloud with Docker

```
1  docker run -d --name nextcloud \
2  -p 443:443 \
3  -v /opt/nextcloud:/var/www/html \
4  nextcloud
```

Security Enhancements:

- Use Let's Encrypt for TLS encryption.

- Enable fail2ban to block brute-force attacks.

- Secure database with PostgreSQL over MariaDB.

19.2 Self-Hosting a Private VPN

A self-hosted VPN prevents your ISP and third parties from monitoring your internet traffic.

Setting up WireGuard VPN (Ubuntu):

```
1    # Install WireGuard
2    sudo apt install wireguard
3
4    # Generate VPN keys
5    wg genkey | tee privatekey | wg pubkey > publickey
6
7    # Enable and start WireGuard
8    sudo systemctl enable wg-quick@wg0
9    sudo systemctl start wg-quick@wg0
```

Security Enhancements:

- Use Mullvad or Tor over VPN for better anonymity.

- Enable firewall rules to block leaks.

- Regularly update WireGuard keys.

Alternative VPNs:

- OpenVPN - More established, supports more devices.

- Streisand - Automatically sets up multiple VPN protocols.

19.3 Running a Matrix Server for Secure Chat

Matrix is a decentralized messaging protocol that ensures full control over private communications.

Setting up a Matrix Homeserver (Synapse):

```
1  # Install Matrix Synapse server
2  sudo apt install matrix-synapse
3
4  # Configure domain settings
5  sudo nano /etc/matrix-synapse/homeserver.yaml
6
7  # Restart Matrix service
8  sudo systemctl restart matrix-synapse
```

Security Enhancements:

- Use Element as a privacy-focused Matrix client.

- Enable E2EE by default for encrypted chats.

- Block unwanted federation for better privacy.

19.4 Dead Drop Networks for Secure Data Exchange

If digital channels are compromised, fallback to physical "dead drop" exchanges.

- **Physical Drop Points:** Pre-agreed hidden locations (e.g., park bench underside) for leaving encrypted drives or messages.

- **Steganography in Physical Objects:** Embed QR codes or micro-SD cards in common objects.

- **Timed Pickups:** Exchange items at randomized intervals to avoid pattern detection.

19.5 Self-Hosting a Privacy-Focused Search Engine (SearXNG)

SearXNG is an open-source, self-hosted search engine that provides Google-like results without tracking.

Installing SearXNG:

```
sudo apt install docker docker-compose
git clone https://github.com/searxng/searxng-docker.git
cd searxng-docker
docker-compose up -d
```

Why self-host SearXNG?

• Avoid Google and Bing tracking.

• Fully customizable search results.

• Can integrate Tor for anonymous queries.

19.6 Setting Up a Private Email Server (Mail-in-a-Box)

Running your own email server improves security, but requires careful setup.

Installing Mail-in-a-Box (Ubuntu 22.04):

```
# Install Mail-in-a-Box
curl -s https://mailinabox.email/setup.sh | sudo bash
```

Security Enhancements:

• Enable PGP encryption for email privacy.

• Use TutaNota or ProtonMail Bridge for encrypted external email.

• Configure DNSSEC to prevent email spoofing.

19.7 Self-Hosting a Tor Hidden Service

Hosting a website or service on the Tor network
ensures maximum anonymity.

Setting up a Tor hidden service:

```
1   # Install Tor
2   sudo apt install tor
3   # Edit torrc configuration
4   sudo nano /etc/tor/torrc
5   # Add hidden service settings
6   HiddenServiceDir /var/lib/tor/hidden_service/
7   HiddenServicePort 80 127.0.0.1:8080
8   # Restart Tor service
9   sudo systemctl restart tor
```

Why host services over Tor?

• Anonymous hosting with hidden IP.

• Resistant to censorship and takedowns.

• Perfect for whistleblowers, activists, and
journalists.

19.8 Anonymous Hosting with Tor Hidden Services

Tor Hidden Services allow you to host Nextcloud or
Syncthing anonymously.

Install Tor:

```
1   sudo apt install tor
```

Configure a Hidden Service:

```
1   sudo nano /etc/tor/torrc
2
3   HiddenServiceDir /var/lib/tor/nextcloud_hidden_service/
4   HiddenServicePort 80 127.0.0.1:8080
```

Restart Tor and Get Your .onion Address:

```
1  sudo systemctl restart tor
2  sudo cat /var/lib/tor/nextcloud_hidden_service/hostname
```

Your Nextcloud is now accessible anonymously over Tor.

19.9 Choosing the Best Self-Hosting Setup

Self-Hosted Service	Best For	Security Features
Nextcloud	Private cloud storage	End-to-end encryption, self-hosted files
WireGuard	Self-hosted VPN	No-logs VPN, strong encryption
Matrix (Synapse)	Private encrypted chat	Federated, E2EE by default
SearXNG	Private search engine	Blocks tracking, no logs
Mail-in-a-Box	Private email server	PGP support, DNSSEC
Tor Hidden Services	Anonymous hosting	No IP exposure, censorship-resistant

Table 19.1: Comparison of Self-Hosting Privacy Services

19.10   Secure Remote Access with Tailscale (Zero Trust VPN)

Tailscale allows you to create a private, encrypted mesh network between your devices without exposing services to the internet.

Install Tailscale on Your Server:

```
1   curl -fsSL https://tailscale.com/install.sh | sh
2   sudo tailscale up
```

Access Your Cloud Securely:

- Access Nextcloud via your private Tailscale IP (e.g., 100.x.x.x).

- No open ports, no public exposure.

- Encrypted, peer-to-peer network access.

19.11   Final Recommendations: Harden Your Self-Hosted Services

To improve the security of your self-hosted tools, apply the following measures: **1. Use Firewalls to Restrict Access**

```
1   # Allow only SSH, HTTP, and HTTPS
2   sudo ufw default deny incoming
3   sudo ufw allow ssh
4   sudo ufw allow http
5   sudo ufw allow https
6   sudo ufw enable
```

2. Enable Fail2Ban to Block Brute-Force Attacks

```
1   sudo apt install fail2ban
2   sudo systemctl enable fail2ban
```

3. Enforce TLS Encryption for All Services

```
1   sudo apt install certbot python3-certbot-nginx
2   sudo certbot --nginx -d yourdomain.com
```

By self-hosting privacy-focused services, you eliminate reliance on corporate surveillance, enhance security, and ensure data sovereignty.

Why This Matters

Every cloud has a provider. *Until you build your own.*
Corporate platforms exist to mine your data.
Regulators move slower than exploit developers.
Self-hosting is more than privacy—it's digital independence. You control the keys, the logs, and the data. No gatekeepers. No backdoors. No apologies.
Welcome to sovereignty on your terms.

20.

OPSEC for High-Risk Individuals

Individuals facing high-risk surveillance scenarios (journalists, activists, whistleblowers, intelligence operatives) must apply strict OPSEC to minimize exposure.

Privacy isn't a preference when your freedom—or your life—is on the line. Whether you're a journalist exposing corruption, a whistleblower leaking government secrets, or an activist under authoritarian watch, you're not paranoid—you're preparing for survival.

Operational Security (OPSEC) is the art of controlling what adversaries know about you. This section is not theoretical. It is tactical, actionable, and tested under real-world pressure.

20.1 Maintaining Anonymity While Traveling

When traveling, adversaries may attempt to track, seize, or compromise your devices. Avoiding digital exposure is crucial.

Best OPSEC Practices for Travel:

- Use burner phones & SIM cards purchased with cash.

- Avoid logging into personal accounts from foreign networks.

- Carry a Tails OS or Qubes OS USB instead of bringing a full laptop.

- Use Faraday bags to prevent remote access to devices.

- Always encrypt data before crossing borders.

- Avoid public Wi-Fi - Use a VPN or travel router with Tor.

Device-Free Travel (Extreme Cases)

- Use a clean laptop with no personal data (air-gapped if possible).

- Carry no personal devices - buy new ones at your destination.

- Travel with a dummy phone containing non-sensitive data.

- Store sensitive files on hidden, encrypted USB drives (VeraCrypt).

Border Crossing Security:

- Store critical data in hidden volumes (deniable encryption).

- Avoid biometric-based travel (e-passports, facial recognition).

- Assume customs may clone your device - wipe before entry.

- Use a disposable phone number in each country.

Field Scenario: Crossing a Hostile Border

You're carrying sensitive leaks on a microSD card. You wipe your laptop to factory state, boot into Tails OS from a hidden USB, and destroy the boot media after upload.
At customs, your device is clean. Your memory is your only password. No traces. No evidence. No leverage against you.

20.2 Avoiding Digital & Physical Surveillance

High-risk individuals must actively evade digital and physical tracking.

How to Prevent Device Seizures:

- Use deniable encryption (hidden VeraCrypt volumes).

- Keep sensitive data on offline, air-gapped devices.

- Store passwords in offline password managers.

- Use encrypted cloud backups (ProtonDrive, Nextcloud).

How to Evade Facial Recognition & AI Surveillance:

- Wear face-obscuring accessories (hats, masks, glasses).

- Use anti-facial recognition makeup or IR LED emitters.

- Avoid ATMs & public cameras with AI tracking.

- Travel at randomized times to avoid pattern detection.

Defending Against AI-Enhanced Surveillance:

- Gait Obfuscation: Change your walking pattern.

- Thermal Imaging Defense: Wear insulated clothing to reduce heat signatures.

- License Plate Covering: Use automated license plate deflectors.

- Drone Evasion: Use high-density urban areas to disrupt drone tracking.

20.3 Using Disposable Identities & Burner Accounts

For maximum anonymity, use compartmentalized burner identities.

How to Create a Secure Burner Identity:

- Use separate email addresses for different tasks.

- Buy prepaid SIM cards with cash.

- Generate fake personas (name, date of birth, address).

- Avoid linking real-life details to online activities.

Best Tools for Anonymous Identities:

- SimpleLogin - Generate burner email addresses.

- VoIP Services (MySudo, JMP.chat) - Anonymous phone numbers.

- Cryptocurrency Mixers - Remove transaction history links.

20.4 Secure Financial Transactions for High-Risk Individuals

Standard banking is fully traceable. Use privacy-enhanced payment methods.

Best Anonymous Payment Methods:

- Monero (XMR) - The most private cryptocurrency.

- Bitcoin via CoinJoin - Breaks transaction links.

- Prepaid Gift Cards - Purchased with cash.

- Barter & Physical Cash - Ultimate anonymity.

Avoid These for High-Risk Transactions:

- Credit/Debit Cards - Fully tracked.

- PayPal/Venmo - Linked to identity.

- Bitcoin (Without Mixing) - Transactions can be traced.

Reality Check

Bitcoin is not anonymous. Every transaction is permanently recorded on a public ledger. Without privacy tools like CoinJoin or privacy coins like Monero, you're leaving a paper trail—just digital.

20.5 Digital Footprint Reduction Strategies

High-risk individuals must erase their digital footprint to prevent tracking.

Data Removal Checklist:

- Request data removal from public records (use services like DeleteMe).

- Delete old social media accounts.

- Use JustDeleteMe to remove online accounts.

- Opt-out of data broker sites (Whitepages, Spokeo, Acxiom).

Compartmentalization of Identities:

- Use different devices and accounts for personal vs. high-risk activities.

- Never use the same username, phone number, or email across platforms.

- Rotate through multiple burner identities.

20.6 Final OPSEC Considerations for High-Risk Individuals

To ensure long-term security:

- Assume all communications are monitored.

- Regularly wipe devices after use.

- Use offline cold storage for important data.

- Never trust unverified contacts – assume social engineering attempts.

- Rotate devices every few months to prevent tracking.

Extreme OPSEC (For Critical Situations)

- Dead Drops – Use physical locations to transfer sensitive data.

- One-Time Pads (OTP) – Unbreakable encryption for critical messages.

- Burner Devices – Use and destroy after high-risk operations.

- Steganography – Hide messages within images/audio files.

By following these OPSEC strategies, high-risk individuals can significantly reduce their exposure to surveillance, tracking, and compromise.

21.

Case Study: Ross Ulbricht - OPSEC Failures

Ross Ulbricht, known as "Dread Pirate Roberts," was the mastermind behind the infamous darknet marketplace, **Silk Road.** While his technical skills were impressive, his **poor OPSEC (Operational Security)** led to his identification, arrest, and life imprisonment.

This case study highlights his critical OPSEC mistakes and provides lessons for avoiding similar pitfalls.

21.1 What He Did Right

- **Tor and Bitcoin**: Silk Road operated as a Tor hidden service, and transactions were conducted in Bitcoin to avoid traditional banking oversight.

- **Compartmentalization (Initially)**: Ulbricht used aliases, encrypted communications, and avoided directly associating his real identity with Silk

Road.

- **Strict Access Control:** Only a few trusted individuals had backend access, reducing insider threats.

21.2 How His OPSEC Mistakes Got Him Caught

The FBI built a digital and physical case against Ulbricht by exploiting the following mistakes.

1. Early Digital Footprint (2011)

Mistake: Used a Personal Email in Public Forums

Ulbricht posted a Tor-related question on Stack Overflow **using his personal Gmail account (rossulbricht@gmail.com).** He later changed his username, but law enforcement had already archived the post. This was a critical error that tied his real identity to Silk Road's backend infrastructure.

Lesson: Always Use Burner Identities

- Use **disposable email addresses** (ProtonMail, Tutanota, SimpleLogin).

- Avoid **real names, locations, or personal identifiers** in public forums.

- Use separate, unlinked aliases for different online activities.

2. Bitcoin Transactions Were Traceable

Mistake: Failed to Properly Anonymize Bitcoin Transactions

Early Silk Road transactions were linked to wallets controlled by Ulbricht. Law enforcement used blockchain analysis to connect these transactions to his real-world accounts.

Lesson: Use Advanced Cryptocurrency Privacy Techniques

- Use Monero (XMR) – Bitcoin is pseudonymous, not anonymous.

- CoinJoin (Samourai Whirlpool, Wasabi Wallet) – Breaks transaction history links.

- Bitcoin Mixers – Avoids direct transaction tracing (beware of honeypot mixers).

3. Sloppy Forum Activity

Mistake: Linked His Real Name to Silk Road on BitcoinTalk (2011)

A user named Altoid promoted Silk Road on BitcoinTalk forums. Later, the same user posted a job listing: *"Looking for a lead developer. Contact me at rossulbricht@gmail.com."*

This provided a direct link between Ulbricht's real identity and Silk Road.

Lesson: Never Cross-Associate Online Identities

- Use completely separate usernames for different activities.

- Avoid reusing writing styles, grammar, or unique phrases.

- Use AI-generated writing patterns to disrupt linguistic profiling.

4. Using the Same Laptop for Personal and Illegal Activities

Mistake: No Device Compartmentalization

When arrested at a public library, Ulbricht was logged into Silk Road as Dread Pirate Roberts. The FBI seized his laptop, which contained:

- PGP private keys

- Silk Road business journal

- Administrative logs and communications

Lesson: Use Dedicated, Compartmentalized Devices

- Use Qubes OS or Tails OS for high-security operations.

- Store critical data on air-gapped systems.

- Implement a dead man's switch to wipe devices remotely.

5. Getting Caught While Logged In

Mistake: Failed to Use Kill-Switches or Session Security

Ulbricht was arrested while logged into Silk Road in a public library. FBI agents staged a distraction to keep him from closing his laptop, allowing them to seize it in an unlocked state.

Lesson: Never Stay Logged In Publicly

- Use hardware kill switches (e.g., Purism laptops).

- Implement dead-man timers to auto-lock/logout when inactive.
- Set up an emergency self-destruct mechanism (USB kill cord, encrypted storage wipe).

21.3 How Law Enforcement Exploited His OPSEC Failures

The FBI used a combination of digital forensics, social engineering, and real-world tactics to capture Ulbricht.

Key Exploits:

- Blockchain forensics - Traced early Bitcoin transactions.
- Metadata analysis - Connected usernames across forums.
- Behavioral pattern analysis - Matched writing styles between accounts.
- Physical surveillance - Staked out public locations where he worked.
- Distraction tactics - Prevented him from locking his laptop.

21.4 Key OPSEC Lessons from Ross Ulbricht

- **Never Use Personal Emails for Anonymous Operations** Always create burner identities and use PGP encryption for communication.
- **Never Link Alias Usernames to Your Real Identity** Even minor online interactions can be correlated by AI-driven investigations.

- **Use Dedicated Devices for Secure Operations** Always compartmentalize by using air-gapped systems or Qubes OS.

- **Anonymize Cryptocurrency Transactions** Bitcoin is traceable. Use Monero, CoinJoin, and mixing services.

- **Never Get Caught While Logged In** Implement dead-man switches, self-destruct triggers, and multi-layer authentication.

- **Physical Security is Just as Important as Digital Security** A single mistake in device handling, travel habits, or surveillance evasion can result in exposure.

Summary: Ross Ulbricht was brilliant, but his OPSEC failures led to his downfall. This case proves that even tech-savvy individuals can be caught due to small mistakes. **Compartmentalization, anonymity, and strict security habits are essential for anyone in high-risk environments.**

22.

Defending Against Nation-State Surveillance

Nation-state actors have access to sophisticated surveillance tools that can track, intercept, and exploit digital devices. Unlike regular cybercriminals, they have vast resources, legal authority, and cooperation from telecom providers. This section outlines countermeasures to defend against state-level surveillance.

22.1 Avoiding IMSI Catchers & Stingray Devices

IMSI Catchers (also called Stingrays or Dirtboxes) are devices that mimic legitimate cell towers, tricking mobile phones into connecting and revealing location, metadata, and even unencrypted calls/SMS.

How IMSI Catchers Work:

- Force phones to downgrade to **2G**, where encryption is weak or nonexistent.

- Intercept and record call/SMS data.
- Perform denial-of-service (DoS) attacks on target devices.
- Inject malicious baseband firmware exploits.

Countermeasures:

- **Use 4G/LTE or 5G Only Mode** – Prevents forced 2G downgrades.
- **Use a "Dumbphone"** – Older 2G-only feature phones are often immune to IMSI catchers.
- **Monitor Cellular Tower Changes:**
  - Use SnoopSnitch (Android) to detect rogue towers.
  - Compare tower signal strength with OpenCellID databases.
- **Keep Your Phone in a Faraday Bag** – Blocks all radio signals when not in use.
- **Use VoIP Over Encrypted Channels** – Instead of direct calls, use Signal, Session, or Briar over VPN/Tor.

22.2 AI-Assisted Surveillance & Metadata Exploitation

Governments increasingly rely on AI and Big Data analytics to track individuals using metadata instead of direct content interception.

How Governments Exploit Metadata:

- Track **call logs and social networks** (who you contact, when, and where).

- Analyze **travel patterns** (geolocation history, routine movements).

- Predict **behavior** (purchasing habits, online activity patterns).

Countermeasures:

- **Use Multiple Identities:**
  - Separate work, personal, and anonymous activities across different devices.
  - Never mix secure communications with normal accounts.

- **Randomize MAC Addresses:**

```
# Enable MAC address randomization
nmcli connection modify <network> wifi.mac-address-
    randomization yes
```

- **Avoid Consistent Schedules:**
  - Change travel routes and routines.
  - Avoid check-ins or social media updates in real-time.

- **Use Cash or Privacy Coins for Purchases:**
  - Avoid credit/debit cards linked to your real identity.
  - Use Monero (XMR) instead of Bitcoin for anonymous transactions.

22.3 Defending Against Advanced Spyware (Pegasus, FinSpy, etc.)

State-sponsored spyware like Pegasus (NSO Group), FinSpy, and Predator can remotely infect smartphones and computers, granting full control to attackers.

How Pegasus and Similar Spyware Work:

• Zero-Click Exploits – No user interaction required; infection occurs via iMessage, WhatsApp, or push notifications.

• Kernel-Level Rootkits – Malware operates at the OS level, making detection difficult.

• Persistent Infection – Survives factory resets and reboots.

Countermeasures:

• **Never Click Suspicious Links:** Most spyware infections come from links in SMS, email, or messaging apps.

• **Use GrapheneOS or Hardened Linux:**
  - GrapheneOS removes remote exploits tied to Google Play services.
  - Linux-based mobile OS options (e.g., PureOS, Ubuntu Touch) reduce attack surface.

• **Regularly Factory Reset High-Risk Devices:**
  - Pegasus infections persist through reboots but not through full disk wipes.

• **Use Burner Devices:**
  - For highly sensitive operations, use disposable devices that are regularly destroyed and replaced.

• **Block Camera/Microphone Access:**
  - Cover laptop and phone cameras with tape or physical shutters.
  - Use hardware kill switches (available on Purism Librem, PinePhone).

22.4 Countering Supply Chain and Hardware-Level Surveillance

Advanced adversaries may compromise devices at the hardware or firmware level before you even receive them.

- **Trusted Hardware Sources:** Avoid ordering critical devices through international or suspicious suppliers.

- **Firmware Verification:** Use tools like Flashrom to verify BIOS/UEFI integrity where possible.

- **Coreboot/Libreboot Devices:** Prefer open-source firmware laptops to reduce closed-source backdoor risks.

- **Physical Tamper Detection:** Apply tamper-evident seals on shipping boxes and hardware casings.

- **Supply Chain Air Gapping:** Sanitize new devices before first use—re-flash firmware and wipe storage.

22.5 Secure Offline Communications

If all digital communication is compromised, alternative methods must be used.

Anonymous Communication Methods:

- Burner Phones & Prepaid SIM Cards
  - Buy phones and SIMs with cash (avoid locations with CCTV).
  - Use each burner device only once, then dispose of it.

- One-Time Pads (OTP) for Secure Messaging

    - For extremely sensitive operations, use handwritten one-time pads for cryptographic communication.

- Dead Drops & Physical Notes

    - Leave information in predetermined locations instead of digital messaging.

- HF/VHF Radios for Off-Grid Comms

    - Shortwave radios can be used for encrypted analog communication.

22.6 Avoiding Facial Recognition & Biometric Tracking

Nation-states deploy AI-driven surveillance networks using facial recognition, gait analysis, and voiceprints.

How Facial Recognition is Used Against You:

- Mass AI-powered CCTV networks match faces to government databases.

- Gait Analysis can track movement patterns even if your face is obscured.

- Voice Recognition systems identify individuals via phone calls and recordings.

Countermeasures:

- **Disrupt AI Recognition:**

    - Wear face-obscuring accessories (masks, sunglasses, scarves).

- Use adversarial AI tools like Fawkes to cloak facial data.

- **Change Physical Appearance Regularly:**

  - Vary clothing, hairstyles, and walking patterns.

- **Disrupt Voiceprint Collection:**

  - Use voice changers for calls.

  - Avoid speaking near smart devices (Alexa, Google Home, Siri).

The Uncomfortable Truth About Nation-State Adversaries

They don't need to "hack" you if they can trick, buy, pressure, or physically grab you. They don't need to break encryption if they can own the supply chain. They don't need malware if they control your telecom provider. They don't need to follow you if your metadata volunteers your every move. Defending against them isn't about tools—it's about awareness, behavior, and discipline. Every choice you make either raises or lowers their cost of success. Raise it until you're not worth the expense.

Summary: Nation-state surveillance is extremely advanced, but not unbeatable. Compartmentalization, anonymity, physical security, and AI disruption are essential for high-risk individuals. Every step towards privacy makes it exponentially harder for adversaries to track or compromise you.

23.

Vehicle & GPS Tracking Evasion

Forget the spy movies. Vehicle tracking isn't fiction—it's already happening to you. Law enforcement, private investigators, corporations, and cybercriminals all exploit vehicle tracking to map your movements, predict your behavior, and expose your associations.

This chapter doesn't just explain how it works. It shows you how to burn their playbook—and make your next move untraceable.

23.1 Understanding Modern Vehicle Tracking Threats

23.2 Common Tracking Technologies

- **GPS Trackers** - Active or passive beacons placed on your vehicle to log or transmit real-time location data.

- **Cellular Triangulation** - Mobile carriers track your phone's tower pings, even without GPS.

- **License Plate Recognition (LPR)** - AI-driven camera systems logging license plates at intersections, highways, and parking lots.

- **Onboard Telemetry** - Modern vehicles transmit speed, braking, and GPS data to manufacturers and rental agencies.

- **Bluetooth Trackers** - Covert AirTags, Tiles, or SmartTags planted on your car or belongings.

- **Vehicle-to-Everything (V2X)** - Emerging systems that broadcast your location to nearby infrastructure.

- **Acoustic Fingerprinting** - AI tracking based on your vehicle's unique engine or exhaust sound.

23.3 Detecting and Removing Tracking Devices

23.4 How to Sweep for Trackers

- **Manual Inspection** - Check under wheel wells, bumpers, dashboards, and engine compartments.

- **Magnetic Wand or Undercarriage Mirror** - Detect hidden magnetic trackers.

- **Thermal Camera Sweep** - Identify battery-powered devices by heat signatures.

- **RF Detector Scan** - Locate active GPS or cellular transmitters.

- **Bluetooth Scanners** - Detect AirTags and similar trackers.

23.5 Detecting Passive Loggers

- Passive trackers do not transmit signals.

- Look for hidden black boxes attached with magnets or zip ties.

- Sweep for unusual objects or tamper marks during vehicle inspections.

23.6 Disrupting Digital Vehicle Telemetry

- **Disconnect Telematics Modules** – Vehicles like Tesla, BMW, and GM have always-on cellular tracking. Physically disconnect SIM modules if possible.

- **Faraday Bags for Key Fobs** – Prevent relay theft and location tracking via passive RF interception.

- **Disable Infotainment Systems** – Factory resets or hardware disconnections can prevent data logging.

- **Avoid Smart Vehicle Rentals** – Use mechanical-key vehicles or trusted private rentals.

- **Avoid Toll Roads and Traffic Cameras** – These systems feed government tracking databases.

23.7 Counter-Surveillance Driving Techniques

23.8 Tactical Driving Methods

- **Three Right Turns Rule** – Make three consecutive right turns to detect tails.

- **Traffic Light Traps** – Force following vehicles to commit or stop.

- **Sudden Route Diversions** – Unexpected turns or exits to flush out surveillance.

- **Vehicle Swap Chains** – Pre-stage cars in secure parking garages to break tracking chains.

- **High-Security Zones** – Drive toward law enforcement or military facilities to deter physical tails.

23.9 Environmental Awareness

- Memorize parked vehicles and note recurring license plates.

- Scan surroundings before and after stops.

- Vary departure times and routes to break patterns.

23.10 Defeating AI-Based Vehicle Recognition

23.11 Vehicle Signature Obfuscation

- **Change Appearance** – Use magnetic panels or wraps to alter color and shape.

- **License Plate Covers or Flippers** – Obstruct or randomize plate visibility where legal.

- **Drive in Dense Traffic or Night Conditions** – Reduce AI system accuracy.

23.12 Acoustic Signature Masking

- **Install Aftermarket Mufflers** – Change exhaust notes.

- **Drive in Noisy Areas** – Mask vehicle signature with ambient sound.

23.13 Advanced Defensive Measures

- **GPS Jammers** – Where legal, disrupt tracking signals.

- **GPS Spoofers** – Feed false location data to trackers.

- **Off-Grid Navigation** – Use offline maps on air-gapped devices to avoid location leaks.

- **Electronic Countermeasures** – Advanced users may employ signal analysis gear to detect digital surveillance.

23.14 Advanced Counter-Tracking Enhancements

Radio Silence Windows (RSW)

- Plan travel through areas where radio signals are naturally blocked—mountain passes, tunnels, or rural dead zones—forcing adversaries to rely on physical tracking.

Covert Swap Points

- Use low-traffic gas stations, underground garages, or rest stops with multiple exits to discreetly change vehicles, drivers, or cargo without being seen.

Thermal Signature Management

- Park near large heat sources (factories, trucks) to obscure your vehicle's infrared signature.

- Avoid idling; excessive heat buildup makes surveillance by IR easier.

Driver Behavioral Masking

- Vary your acceleration, braking, and cornering habits regularly.

- Rotate drivers if possible to confuse AI-based driver profiling.

Blended Convoy Operations

- Travel with multiple trusted vehicles to obscure which car contains the actual target.

- Split off in different directions to mislead tracking efforts.

Reversed Surveillance (Counter-Tail Recon)

- Use dashcams or covert cameras to record trailing vehicles for later analysis.

- Look for recurring patterns: same make/model, behavior near turns or stops.

Travel Noise Discipline

- Never discuss routes or meetings over phone or apps.

• Coordinate critical moves in person or via offline encrypted notes.

Non-Digital Navigation Discipline

• Use physical maps or air-gapped GPS devices.

• Avoid smartphone-based navigation apps that leak location data.

The Multi-Dimensional Reality of Vehicle Tracking

GPS tracking is only the beginning. Advanced surveillance integrates:

• GPS and cellular triangulation

• LPR and AI-assisted video tracking

• Acoustic fingerprinting

• Telematics and rental car telemetry

• Bluetooth and RF beacons

24.

AI-Based Cyber Threats

24.1 Surveillance Isn't New. It's Just Faster Now.

In 1970, a young Army intelligence officer named Christopher Pyle blew the whistle on what he called "the Army's domestic surveillance machine."

The U.S. military, you see, wasn't just busy watching foreign threats—they were running a nationwide surveillance dragnet on millions of ordinary Americans. Protesters. Clergy. Students. Politicians. Anyone who dared speak too loudly, too often, or too disruptively.

And they didn't need AI to do it.

They used file cabinets, index cards, and typewriters. Field agents in trench coats—not Python scripts—watched from the shadows. Tens of thousands of names were meticulously cataloged, cross-referenced, and flagged as "subversive."

Pyle didn't just leak this; he testified before Congress. His revelation sparked a massive public outcry, forcing investigations and new laws aimed at reining in the intelligence community.

Spoiler: It didn't last.

You see, they didn't stop. They upgraded.

The surveillance state learned from its exposure. It traded field agents for algorithms. Index cards for data lakes. Trench coats for tracking pixels.

While you were busy setting your Instagram bio, they were perfecting metadata profiling. While you argued over which VPN was better, they were integrating AI-driven behavioral models into telecom infrastructure.

They have a fifty-year head start.

This is the part where you might feel like it's already too late. It isn't. But only if you stop playing catch-up and start playing dirty.

In the next section, you'll learn how to dismantle your digital fingerprint, break their profiling engines, and turn their head start into dead weight.

Let's make Christopher Pyle proud.

Artificial Intelligence (AI) has revolutionized cybersecurity—both for defenders and attackers. Threat actors now leverage AI for automated hacking, deepfake social engineering, AI-assisted surveillance, and real-time deception tactics.

24.2 Deepfake Attacks: AI-Powered Social Engineering

Deepfake technology uses neural networks (GANs - Generative Adversarial Networks) to create hyper-realistic fake audio, video, and images. This is increasingly used for:

- Use a secure browser like Tor Browser or LibreWolf.

- Enable full-disk encryption on all devices.

Countermeasures:

- **Liveness Detection:** Verify video calls using challenge-response authentication (real-time head movements, hand gestures).

- **Multi-Factor Identity Verification:** Avoid relying on voice or facial recognition alone; combine with cryptographic authentication (PGP, FIDO2).

- **Deepfake Detection Tools:** Use AI-powered detection platforms like Microsoft Video Authenticator, Sentinel AI, or Deepware Scanner.

- **Behavioral Biometrics:** Instead of traditional authentication, use keystroke dynamics, typing rhythm, and mouse movement patterns.

AI-Generated Social Engineering at Scale

AI is now capable of generating highly convincing phishing campaigns, voice impersonations, and fake social media profiles with near-human accuracy.

Emerging Threats:

- **GPT-Style Phishing Engines:** Automated phishing that personalizes messages in real time.

- **AI Voice Cloning:** Perfectly mimicked phone calls used for CEO fraud or ransom scams.

- **Synthetic Relationship Building:** Bots that establish long-term trust before launching an attack.

Countermeasures:

- Train teams to spot **too-perfect language or rapid-response patterns.**

- Require **multi-factor or verbal confirmation** for all financial or sensitive requests.

- Deploy **internal-only secure channels** for high-trust communication.

24.3 AI-Assisted Surveillance & Tracking

Governments and threat actors increasingly use AI for real-time surveillance, metadata analysis, and behavioral prediction.

How AI Surveillance Works:

- Facial Recognition – AI-powered systems like **Clearview AI** match faces to massive databases.

- Behavioral Prediction – AI correlates movement, social media activity, and purchase history to predict behavior.

- Gait Analysis – AI can identify individuals even if their face is obscured by analyzing how they walk.

- Emotion Detection – AI-powered CCTV cameras track facial expressions to detect stress, fear, or deception.

Countermeasures:

- Anti-Facial Recognition Techniques:
  - Use CV Dazzle (Adversarial Makeup) to confuse AI.
  - Wear infrared LED glasses to disrupt facial tracking.
  - Apply Fawkes or Glaze to alter digital images and prevent AI profiling.
- Defeating Gait Analysis:
  - Change your walking pattern periodically.
  - Use weighted shoes or attach subtle movement disruptors.
- Metadata Obfuscation:
  - Use VPNs, Tor, and MAC address spoofing.
  - Randomize online activity times and locations.

24.4 Behavioral Surveillance and Stylometric Profiling Defense

Your VPN is active. Your Tor Browser is configured. You're cloaked, or so you think. But there's something you forgot... **you**.

The greatest vulnerability isn't your IP. It's your **behavior**.

24.5 The Hidden Threat of Stylometry

Stylometry is the forensic analysis of writing style—sentence structure, vocabulary, punctuation, and even typographical quirks. Law enforcement has used it to deanonymize whistleblowers and activists. Big Tech uses it to correlate identities across platforms.

Every tweet. Every post. Every comment. You leave a signature.

How to Fight Stylometric Profiling

- Vary your writing style intentionally across different identities.

- Use AI rewriters or paraphrasers to break consistent writing patterns.

- Avoid posting identical content across multiple platforms.

- Mix sentence lengths, vocabulary, and tone.

24.6 Keystroke and Mouse Behavior Tracking

You are also profiled by how you type. Keystroke cadence, typing speed, mouse movements—these are biometric signals silently logged by surveillance systems.

Countermeasures

- Use virtual keyboards or touch interfaces when possible.

- Employ mouse randomizer tools to break movement patterns.

- Limit real-time interaction on high-risk platforms.

24.7 Voiceprint and Audio Fingerprinting

Voice is a biometric signature. Virtual assistants like Siri, Alexa, and Google Assistant collect this data by design. Voice recognition is increasingly used by state-level surveillance programs.

Countermeasures

- Avoid using voice-controlled features on any device.

- Use hardware voice changers or software-based voice masking tools.

- Physically disable or tape over microphones when not in use.

24.8 Hardware Supply Chain and Firmware Integrity Defense

You ordered your laptop online. It arrived sealed and flawless. But can you trust it?

24.9 Beyond Intel ME: The Silicon You Can't See

Nearly all modern processors ship with hidden subsystems—Intel Management Engine (IME), AMD Platform Security Processor (PSP), or Apple's Secure Enclave. These are black boxes with full hardware-level access.

Why You Should Care

- They operate below your operating system, invisible to antivirus and security tools.

- They are closed-source and unauditable.

- They support remote attestation, turning your device into a silent informant.

24.10 Preferred Open Hardware Platforms

- **Purism Librem Series** - Coreboot firmware, hardware kill switches.

- **System76 Open Firmware** - Coreboot-powered, open-source stack.

- **Framework Laptop** - Repairable and increasingly open-source.

24.11 DIY Anti-Interdiction for Everyday Buyers

- Photograph packaging, serial numbers, and seals before opening.

- Document timestamps, delivery details, and unboxing process.

- Reflash firmware or BIOS immediately after receiving new hardware.

24.12 Verifying Firmware Integrity

- Use `flashrom` to read and verify firmware hashes.

- If supported, install coreboot or Heads for tamper-evident boot.

- Validate firmware signatures when available from trusted vendors.

24.13 Establishing a Community Hardware Trust Network

- Collaborate with trusted peers to cross-verify hardware.

- Share validated hardware serials and hashes in community registries.

- Build decentralized trust through collective validation.

Trust nothing sealed in plastic. Verify everything. You are the final auditor of your security.

24.14 AI-Enhanced Multi-Vector Campaigns (Attack Chain Automation)

AI is no longer confined to a single tactic. Modern threat actors chain AI-powered techniques across the entire kill chain.

Example Campaign Flow:

1. AI scans for exposed credentials in OSINT and breach data.

2. AI-generated phishing emails or voice clones target the user for MFA bypass.

3. AI identifies weak points in the target's infrastructure or supply chain.

4. AI generates real-time polymorphic payloads tailored to the discovered vulnerabilities.

5. AI exfiltrates data while manipulating logs to evade detection.

Countermeasures:

- Monitor entire attack surfaces, not just endpoints.

- Conduct adversary emulation exercises to simulate AI-powered attack chains.

- Use deception and honeypots to detect multi-stage campaigns early.

24.15 AI-Powered Cyber Crime-as-a-Service (CaaS)

AI has lowered the barrier to entry for cybercriminals. Script kiddies can now rent AI-driven hacking tools, deepfake services, and autonomous phishing engines.

Real-World Examples:

- AI-powered phishing kits sold on dark web forums.

- Ransomware-as-a-Service with AI-generated extortion emails.

- Deepfake voice rental services used in executive fraud scams.

Countermeasures:

- Monitor dark web marketplaces for mentions of your organization.

- Train users to recognize high-quality social engineering attempts.

- Leverage threat intelligence feeds focused on Cyber Crime-as-a-Service ecosystems.

24.16 AI-Augmented Autonomous Reconnaissance Drones

AI-powered drones can conduct physical reconnaissance, RF scanning, and camera surveillance without human operators.

Emerging Threat Vectors:

- AI-guided drones mapping facility perimeters.

- RF sniffing for unsecured Wi-Fi or Bluetooth devices.

- Capturing video or IR imagery for surveillance or blackmail.

Countermeasures:

- Use drone detection systems (RF, radar, or acoustic).

- Physically shield sensitive operations from line-of-sight observation.

- Enforce strict no-drone zones and physical surveillance countermeasures.

AI-Powered Behavioral Fingerprinting Evasion

AI builds behavioral profiles based on online habits, movement patterns, and interaction styles.

Behavioral Fingerprinting Vectors:

- Typing rhythm, mouse movement, and click patterns.

- Online behavior timelines (when and how you interact).

- Physical movement patterns in smart environments.

Countermeasures:

- Vary interaction methods—use different devices, typing styles, and schedules.

- Deploy privacy tools that inject randomized behavior (e.g., mouse jitter tools).

- Avoid consistent device usage patterns across multiple identities.

24.17 AI-Powered Physical Security Breaches

AI isn't just limited to cyberspace. Physical security systems are increasingly vulnerable to AI-assisted attacks.

Examples of AI-Powered Physical Breaches:

- AI-powered lock-picking robots that analyze lock feedback in real time.

- AI-assisted CCTV analysis used to identify gaps in patrol or camera coverage.

- Machine-learning-based alarm evasion, where AI learns sensor patterns and adapts attacks to avoid detection.

Countermeasures:

- Use randomized patrol routes and camera sweeps.

- Implement sensor fusion—combine thermal, motion, and vibration sensors to avoid single-point failures.

- Deploy AI-driven counter-surveillance tools to detect pattern analysis attempts.

Weaponized AI for Automated Reconnaissance

AI is now used by threat actors to accelerate reconnaissance and target profiling.

How AI is Used for Recon:

- Scrapes public profiles, forums, and repositories to build detailed target dossiers.

- Uses NLP (Natural Language Processing) to extract emotional states, schedules, and network maps from open-source data.

- Identifies exploitable technologies via automated banner grabbing and Shodan-like AI-enhanced scanning.

Countermeasures:

- Regularly scrub personal and corporate data from OSINT platforms.

- Obfuscate technology stacks and server fingerprints (e.g., modify HTTP headers).

- Use deception technologies (honeypots, honeyports) to feed false data to automated recon bots.

24.18 AI-Driven Legal Exploitation (Lawfare Tactics)

Adversaries can weaponize AI-powered legal discovery, patent trolling, and compliance weaponization to drain resources or damage reputations.

Examples:

- AI scanning of patent databases to identify lawsuits with financial leverage.

- AI drafting mass legal claims, takedown notices, or compliance threats.

- AI targeting regulatory reporting to create reputational damage.

Countermeasures:

- Legal teams should monitor AI-generated threats or abuse of legal frameworks.

- Use AI-powered legal research tools to defend against lawfare.

- Maintain legal response playbooks for rapid response to weaponized legal tactics.

24.19 AI-Augmented Supply Chain Attacks

AI is increasingly used to automate the discovery and exploitation of supply chain weaknesses.

Examples of AI-Enhanced Supply Chain Threats:

- AI-powered scanning of vendor ecosystems to identify weak links.

- Automated analysis of software dependencies to inject malicious packages.

- AI-assisted phishing targeting suppliers or partners to reach your organization.

Countermeasures:

- Perform continuous **supply chain risk assessments.**

- Monitor for **dependency confusion** or typosquatting in software packages.

- Require **vendor security audits** and supply chain visibility reports.

- Harden vendor access with **least privilege and segmentation.**

AI-Enhanced Insider Threats

AI-powered tools are now in the hands of disgruntled employees or insider agents.

Emerging Insider Threat Techniques:

- Using AI to auto-classify sensitive files for exfiltration.

- Auto-generating phishing payloads or malware specifically tailored to internal tools.

- Bypassing security training through AI-generated fake compliance responses.

Countermeasures:

- Deploy User and Entity Behavior Analytics (UEBA) to flag anomalies.

- Implement strict data access segmentation—need-to-know only.

- Use canary tokens (fake sensitive files) to detect internal reconnaissance.

24.20 AI-Assisted Social Engineering at Scale

Attackers now use AI to personalize scams at a massive scale—no longer relying on poorly-written phishing emails.

Examples of AI-Driven Social Engineering:

- AI that learns corporate hierarchy to impersonate executives in spear-phishing attacks.

- Automated voice cloning for CEO fraud or "family emergency" scams.

- Machine-generated LinkedIn or social media profiles used to establish trust before attacks.

Countermeasures:

- Implement strict verbal verification procedures for sensitive requests.

- Train staff to detect synthetic media (voice or video).

- Use internal-only communication channels with cryptographic verification.

Cognitive Warfare & Psychological Manipulation

AI is used to weaponize human psychology through fear, uncertainty, and manipulation.

- **Disinformation Fatigue Management:** Avoid engaging with endless conflicting narratives—verify, then disengage.

- **Emotional Trigger Awareness:** Recognize attempts to generate fear, outrage, or urgency.

- **Critical Thinking as a Defense:** Apply source verification, fact-checking, and skepticism to all digital content.

24.21 AI-Powered Hacking & Autonomous Exploits

Cybercriminals are now deploying AI-powered attack frameworks to automate hacking.

Examples of AI-Driven Cyberattacks:

- AI-Generated Malware: Polymorphic malware changes its code in real-time to evade detection.

- AI-Powered Password Guessing: AI cracks passwords by analyzing social media habits.

- Automated Phishing Attacks: GPT-like AI generates highly convincing, personalized phishing emails.

Countermeasures:

- Use **behavior-based detection systems** (EDR/XDR solutions) instead of signature-based antivirus.

- Implement hardware security keys (YubiKey, Titan Security Key) to prevent AI-driven credential stuffing.
- Train employees to recognize AI-generated phishing emails.

AI-Driven Zero-Click Exploitation

AI can identify and weaponize zero-click vulnerabilities, exploits requiring no user interaction.

Examples:

- Malicious image, PDF, or media file rendering bugs.
- Exploits triggered by receiving a message, email, or call without opening it.

Defensive Measures:

- Disable automatic media and document previews in apps.
- Apply patches and updates promptly.
- Use **exploit mitigation tools** (e.g., AppArmor, Windows Exploit Protection).

24.22 Polymorphic Malware and Autonomous Exploitation

AI-driven malware no longer relies on static payloads. It evolves in real-time, rewriting itself to evade traditional defenses.

Examples of AI-Enhanced Threats:

- **Polymorphic Malware:** Malware that mutates its code or behavior to bypass antivirus detection.

- **Autonomous Exploitation:** AI systems that scan, identify, and exploit vulnerabilities without human oversight.

- **Live Payload Generation:** AI that custom-builds malware based on the target's OS, apps, or network stack.

Countermeasures:

- Deploy **Behavior-Based Detection** (EDR/XDR) that focuses on malicious actions, not just signatures.

- Monitor for **Process Injection, Memory Tampering, and Unusual Network Activity.**

- Use **Honeypots and Canary Files** to detect autonomous malware behavior.

24.23 Code Signing Abuse to Evade Detection

Malware may use stolen or legitimate certificates to appear trusted.

Examples:

- Signed malware bypassing antivirus software.

- Exploiting trust in **digitally signed binaries.**

Defensive Measures:

- Validate **certificate chains and revocation lists.**

- Use **behavior-based detection** over signature trust alone.

Adversarial AI Exploitation and Model Poisoning

Attackers don't just use AI—they target the AI itself.

How AI Models Get Compromised:

- **Data Poisoning:** Injecting false data into AI training sets to bias or break models.

- **Evasion Attacks:** Crafting inputs designed to bypass AI detection.

- **Model Inversion:** Extracting sensitive data from deployed AI models.

Countermeasures:

- Apply **Differential Privacy** to protect training data.

- Use **Adversarial Training** to harden models against crafted attacks.

- Monitor AI behavior with **AI-Watching-AI** systems to detect abuse.

AI-Augmented Disinformation Campaigns

State-sponsored actors use AI to spread propaganda, social engineering, and automated bot networks.

How AI Enables Disinformation:

- AI generates fake news articles and mass-produces bot accounts.

- Chatbots like Tay AI (Microsoft) were manipulated to spread hate speech.

- AI deepfake videos fuel political instability.

Countermeasures:

- Cross-check sources using OSINT fact-checking tools (Hoaxy, InVID, NewsGuard).

- Monitor for AI-generated text using GPTZero or Deepfake Text Detectors.

- Disable automated bot amplification via browser extensions (Bot Sentinel).

24.24 Adversarial AI Attacks: Hacking AI Itself

Hackers now target AI systems by feeding them manipulated data (known as adversarial attacks).

How Adversarial Attacks Work:

- Poisoning AI Datasets – Injecting manipulated training data into AI models.

- Evasion Attacks – Modifying input data to deceive AI (e.g., making self-driving cars ignore stop signs).

- Model Inversion – Extracting private training data from AI models.

Countermeasures:

- Use differential privacy to prevent AI models from leaking sensitive training data.

- Apply adversarial training to harden AI against evasion attacks.

- Deploy AI anomaly detection to monitor AI behavior for manipulation.

Summary: AI-powered cyber threats are rapidly evolving, making traditional security measures obsolete. Defending against AI-based attacks requires **behavioral security, adversarial AI countermeasures,** and **privacy-first digital strategies.**

25.

Facial Recognition & Gait Analysis Evasion

25.1 Understanding Facial Recognition Threats

Facial recognition technology is widely used for surveillance, law enforcement, and commercial tracking. AI-driven facial recognition systems analyze facial features, comparing them with large datasets to identify individuals.

Key threats:

- Government surveillance programs (e.g., Clearview AI, PimEyes)

- Retailer and public-space tracking

- Automated border security and law enforcement databases

25.2 Techniques for Evasion

To counter facial recognition, consider the following:

- **Adversarial Fashion** - Wearing patterns that confuse AI algorithms (e.g., CV Dazzle makeup techniques)

- **Infrared Interference** - Glasses with embedded IR LEDs can disrupt camera sensors

- **Obscuring Facial Features** - Masks, hats, and sunglasses prevent complete facial scans

- **Deepfake Overlays** - Using real-time digital face masking to manipulate recognition

25.3 Gait Recognition & Motion Analysis

Many surveillance systems now analyze a person's unique walking pattern, even when their face is obscured.

Countermeasures:

- **Posture Alteration** - Slightly altering walking patterns can disrupt AI tracking

- **Weighted Clothing** - Changes movement dynamics, confusing gait recognition

- **Randomization** - Avoiding predictable movement patterns

25.4 Evasion Through Environmental Exploit

Sometimes the best evasion isn't technological—it's environmental. Use surroundings to defeat camera angles and sensor capabilities.

- **Camera Blind Spots:** Walk along walls or use physical structures like poles, signs, or parked vehicles to obscure line-of-sight.

- **Backlighting Techniques:** Walk with bright lights or sunlight directly behind you to reduce facial scan accuracy.

- **Crowd Blending:** Move within dense groups of people to overload tracking systems with multiple potential matches.

25.5 Active Anti-Surveillance Wearables

Some specialized wearables are designed to disrupt AI surveillance in real time.

- **IR-LED Hats or Scarves:** Emit invisible infrared light to blind near-IR cameras used in night vision or security systems.

- **Reflective Fabrics:** Specialized materials that bounce back camera flashes or LIDAR pulses, confusing 3D facial scanners.

- **Wearable Signal Jammers:** (Where legal) Portable RF or IR jamming devices to interfere with sensor systems.

25.6 Cross-Device Correlation Attacks

You may evade facial or gait recognition in public, but cross-device data correlation can still re-identify you.

Example Threats:

- Mobile phones, smartwatches, earbuds, or Bluetooth beacons broadcasting identifying signals (MAC addresses, device fingerprints).

- Correlation between your physical movement and your online activity or device connections (e.g., logging into Wi-Fi at a nearby cafe while evading cameras).

- Wearable sensors (e.g., fitness trackers) leaking motion or health data.

Countermeasures:

- Power down or use Faraday bags for all personal electronics during movement.

- Use devices with randomized MAC addressing and minimal RF emissions.

- Avoid carrying personal electronics when practicing physical evasion.

25.7 Thermal Imaging and Infrared Tracking Evasion

Some surveillance systems deploy thermal cameras or infrared sensors that can detect body heat or IR reflections even in complete darkness.

Emerging Threat Vectors:

- Body heat detection through walls or in low-light conditions.

- IR motion tracking in public spaces (common in smart buildings).

- Thermal residue on objects (e.g., door handles, keyboards).

Countermeasures:

- Use thermal-blocking or reflective clothing to mask heat signatures.

- Avoid lingering near objects you interact with to reduce thermal residue detection.

- Exploit environmental heat sources (e.g., standing near running engines, HVAC vents) to mask your thermal profile.

25.8 Adversarial Object Carrying

Carrying non-obvious props or objects can interfere with AI classification systems.

- **Oversized Objects:** Carrying items like large bags, umbrellas, or boxes can obstruct body posture and gait.

- **Symmetry Disruption:** Wearing asymmetrical clothing or gear (e.g., one weighted shoe) to break pattern recognition.

25.9 Psychological Warfare: Poisoning the Dataset Before They Find You

Why play defense when you can play offense? Poisoning public datasets makes you harder to catalog before surveillance systems even start watching.

- **Fawkes AI Cloaking:** Pre-process your online photos to make them unusable by facial recognition engines without degrading human-viewed quality.

- **LowKey AI Defense:** Leverages adversarial machine learning to permanently alter how AI models perceive your facial features.

- **Mass Disruption Campaigns:** Community-led projects designed to flood facial recognition datasets with poisoned or adversarial samples, reducing model accuracy at scale.

25.10 Environmental and Optical Disruption Tactics

Environments can be weaponized to break surveillance lock-on.

- **Reflective and High-Contrast Environments:** Seek out areas with glass, mirrors, or strong light reflections to disrupt camera sensors.

- **Strobe or Flashlight Blinding:** Small high-intensity strobes or flashlights can oversaturate camera sensors temporarily (use carefully in legal contexts).

- **Environmental Noise Generation:** High foot traffic or chaotic environments can overwhelm motion tracking systems.

Note on Acoustic and Sensor-Based Tracking: Surveillance isn't limited to visual input. Microphones, lidar, and ultrasonic sensors are increasingly used to map environments or track individuals by voice, footsteps, or ambient noise. Countermeasures include white noise generators, vibration dampening materials, and sound masking devices to disrupt acoustic profiling or eavesdropping systems.

25.11 Behavioral Misdirection & Social Engineering Evasion

Sometimes the best way to evade surveillance isn't to hide—but to *mislead*.

- **Decoy Identity Projection:** Wear clothing, accessories, or branding associated with high-profile individuals, corporate employees, or recognizable public figures to create false identity matches in surveillance systems.

- **Pattern Confusion with Coordinated Movements:** Move in pairs or small groups where participants deliberately swap positions, adjust pace, or mimic each other's movements. This creates conflicting signals and identity collisions in AI tracking systems.

- **Environmental Distraction:** Time your movements with crowded environments, visual noise, or high-motion areas to overwhelm tracking algorithms and blend into the chaos.

- **Confidence as Camouflage:** Adopt the body language, posture, or pace of someone who *belongs*. Surveillance systems—and the humans monitoring them—are less likely to flag what looks routine. You don't have to disappear—you just have to make them doubt what they're seeing.

Legal Levers Against Biometric Surveillance

While most of this chapter focuses on technical and behavioral evasion, you may also have **legal rights** depending on your location.

- **Data Removal Requests:** Under laws like the **GDPR** (Europe) and **BIPA** (Illinois, USA), you may have the right to demand the deletion of biometric data collected without your consent.

- **Transparency Demands:** Some jurisdictions require companies to disclose the existence and purpose of facial recognition systems. You can request this information to expose hidden surveillance.

- **Legal Action or Complaints:** Privacy advocacy groups or data protection agencies (like the **ICO** in the UK or the **EDPB** in the EU) may accept complaints or support legal challenges against unauthorized biometric tracking.

- **Public Advocacy:** In regions lacking strong privacy laws, organized pressure on lawmakers and companies can sometimes lead to policy changes or voluntary rollbacks.

Disclaimer: Legal rights vary drastically by jurisdiction. Always consult a qualified privacy or data protection attorney before pursuing legal action.

26.

Cryptocurrency Privacy & Financial Anonymity

Many assume that cryptocurrency is private by default, but most blockchains—such as Bitcoin and Ethereum—are fully traceable. This section outlines practical strategies for financial anonymity, blockchain obfuscation techniques, and privacy-preserving transaction methods.

26.1 Bitcoin Privacy: Avoiding Blockchain Analysis

Bitcoin transactions are permanently recorded on a public ledger, making them vulnerable to blockchain forensics firms (Chainalysis, Elliptic, CipherTrace). Without proper precautions, exchanges, governments, and law enforcement can track your financial history.

How to Improve Bitcoin Privacy:

- **Avoid Reusing Bitcoin Addresses** – Always generate a new receiving address for every transaction to prevent address clustering.

- **Use CoinJoin Mixers** - Privacy-focused wallets like Whirlpool (Samourai Wallet), Wasabi Wallet, and JoinMarket break transaction history links.

- **Run Your Own Bitcoin Node** - Using your own node (Bitcoin Core, Electrum Personal Server) prevents wallet providers from tracking your transactions.

- **Use Lightning Network for Private Transactions** - The Lightning Network enables off-chain, pseudonymous payments that are not recorded on the Bitcoin blockchain.

Using Whirlpool CoinJoin for Bitcoin Anonymity:

```
1   # Run Samourai Wallet's Whirlpool for Bitcoin mixing
2   whirlpool-client-cli --cli --server mainnet
```

Monero (XMR): The Only Truly Private Cryptocurrency

Unlike Bitcoin, Monero (XMR) is private by design and does not require additional obfuscation tools.

Why Monero is Superior for Privacy:

- RingCT (Ring Confidential Transactions): Hides sender, receiver, and transaction amount.

- Stealth Addresses: Prevent blockchain observers from linking transactions to a public address.

- No Traceable Balances: Unlike Bitcoin, Monero does not expose wallet balances or history.

Installing a Monero Wallet for Private Transactions:

```
1   # Download Monero CLI Wallet
2   wget https://downloads.getmonero.org/cli/linux64
3
4   # Extract and run
5   tar -xvf monero-linux-x64.tar.bz2
6   ./monero-wallet-cli
```

Buying Monero Anonymously (No KYC)

To obtain Monero without linking it to your identity, use peer-to-peer (P2P) exchanges:

- LocalMonero - Decentralized P2P exchange for buying Monero with cash or crypto.

- Haveno - A Bisq-based, privacy-focused Monero exchange.

- Bisq - P2P Bitcoin exchange that allows private Monero trades.

26.2 Privacy Coins vs. Bitcoin Mixers: Which is Safer?

Privacy Method	Advantages	Disadvantages
CoinJoin (Bitcoin Mixers)	Decentralized, non-custodial mixing	Requires users to trust that CoinJoin participants are not surveilled
Lightning Network	Off-chain transactions, harder to track	Still requires an entry/exit point into the Bitcoin blockchain
Monero (XMR)	Fully anonymous by default, no extra mixing needed	Not widely accepted on major exchanges due to regulatory pressure

Table 26.1: Comparison of Privacy Techniques in Cryptocurrency

26.3 Avoiding Common Mistakes in Crypto Privacy

Even when using privacy coins, operational security (OPSEC) failures can expose your identity.

How to Prevent Tracing:

- **NEVER withdraw Bitcoin or Monero directly to a real-name bank account.**
- Avoid KYC exchanges like Coinbase, Binance, or Kraken, which link transactions to your identity.
- Use Tor or a VPN when accessing wallets or exchanges to prevent IP tracking.
- Store cryptocurrency in air-gapped hardware wallets (Trezor, Coldcard) for maximum security.
- Use stealth addresses when receiving Monero to prevent linking transactions to a single identity.

Monero Remote Node Privacy Considerations

Using public Monero nodes may expose your IP address and transaction requests.

- **Use Your Own Node:** Always run your own Monero full node if possible.
- **Tor Hidden Service Nodes:** If using remote nodes, connect to .onion nodes over Tor for privacy.

26.4 Advanced Financial Privacy Techniques

To remain anonymous in crypto transactions:

1. **Use Disposable Wallets:**

- Always create a new wallet for high-risk transactions.
- Use Electrum (for Bitcoin) or Monero CLI to generate temporary wallets.

2. Avoid Dust Attacks:

- Dust attacks involve sending tiny amounts of Bitcoin to track wallets.

- Regularly consolidate or move small UTXOs to avoid tracking.

3. Conduct P2P Cash-for-Crypto Swaps:

- Meet in public places for cash transactions.

- Use disposable burner phones and ProtonMail for negotiations.

Cross-Chain Swapping Without KYC

Cross-chain atomic swaps allow you to exchange Bitcoin for Monero directly without relying on custodial exchanges.

- **Haveno** – Decentralized Monero-to-Bitcoin swapping based on Bisq's architecture.

- **Atomic Swaps** – Trustless swaps using smart contracts (see COMIT Network or Farcaster projects).

- **Avoid Custodial Swappers** – Always prefer decentralized, non-custodial swap services to prevent theft or surveillance.

Example: An Anonymous Crypto Purchase (Best OPSEC Practices)

To acquire Bitcoin and convert it into Monero without linking your real identity:

1. Buy Bitcoin Anonymously

   - Use Paxful or Bisq to buy Bitcoin with cash (no KYC).
   - Use a new Bitcoin wallet address (never reuse addresses).

2. Use CoinJoin to Break the Trail

   - Use Wasabi Wallet or Samourai Wallet's Whirlpool to mix Bitcoin.

3. Convert to Monero (XMR)

   - Send mixed Bitcoin to a non-KYC exchange (FixedFloat, KSwap, or MorphToken) to convert BTC to XMR.
   - Withdraw Monero to a newly generated Monero wallet address.

4. Use Monero for Transactions

   - Monero transactions are private by default, hiding sender, receiver, and amounts.

5. Store Crypto Securely

   - Store Monero in an air-gapped cold wallet (Trezor, Coldcard, paper wallet).
   - Never expose wallet keys online.

26.5 Avoiding Blockchain Dust & Metadata Leaks

Beyond dust attacks, metadata leakage can still occur through wallet software or careless use.

- **Turn Off Telemetry:** Disable any data-sharing or telemetry options in your wallet software.

- **Avoid Web Wallets:** Browser-based wallets often leak more metadata than you realize.

- **Validate Wallet Fingerprints:** Always verify wallet binaries and PGP signatures to avoid supply-chain attacks.

26.6 Legal and Regulatory Caution

Many privacy techniques are restricted or illegal in certain jurisdictions.

- **Mixers Banned:** Countries like the U.S. have sanctioned services like Tornado Cash.

- **KYC Enforcement:** Some regions mandate identity verification for all crypto exchanges.

- **Local Laws Vary:** Understand the legal implications in your jurisdiction before using privacy-enhancing techniques.

Summary: Most cryptocurrencies are not private by default. To ensure true anonymity, use Monero (XMR), Bitcoin mixing (CoinJoin), Lightning Network, and avoid KYC exchanges. Always practice strong OPSEC to prevent blockchain analysis from linking transactions to your identity.

27.

OSINT Awareness & Digital Footprint Reduction

Open-Source Intelligence (OSINT) involves gathering information from publicly available sources to profile individuals, organizations, and infrastructure. Attackers, hackers, law enforcement, and corporations use OSINT to track, de-anonymize, and exploit digital footprints.

This section covers how OSINT is conducted, methods attackers use, and how to eliminate personal data leaks.

27.1 How Attackers Gather Information (OSINT Techniques)

OSINT threats come from multiple sources:

1. Social Media Intelligence (SOCMINT)

- Analyzing social media activity to determine habits, interests, and locations.
- Using username tracking across multiple

platforms to correlate identities.

- Extracting metadata from uploaded images (EXIF data) to find GPS locations.

2. Data Breach Exploitation

- Searching leaked credentials on HaveIBeenPwned, Dehashed, and IntelX.

- Cross-referencing email addresses in past breaches to build identity profiles.

- Using the dark web to buy or trade stolen information.

3. Google Dorking & Search Engine Hacking

- Google Dorking finds sensitive data using advanced search operators:

```
1    site:example.com filetype:pdf "confidential"
2    intitle:"index of" site:example.com
3    inurl:/wp-content/uploads/ site:example.com
4    site:pastebin.com intext:"password"
```

- Crawling public archives, pastebins, and forums for exposed data.

- Leveraging reverse image searches (Google, Yandex, PimEyes) to link profiles.

4. Domain & Infrastructure OSINT

- Using WHOIS lookups to identify domain registrants.

- Conducting DNS enumeration to uncover subdomains and hidden services.

- Scanning for leaked admin panels and login portals.

5. AI-Assisted OSINT (Facial Recognition, Pattern Analysis)

- AI-driven tools (Clearview AI, PimEyes) match faces across multiple images.

- Behavioral tracking analyzes posting habits, writing style, and location metadata.

- Cross-referencing dark web and surface web accounts to find linked identities.

27.2 Removing Personal Data from the Internet

Reducing your digital footprint requires active monitoring and deletion of personal information.

Step 1: Delete Unused Accounts

- Use JustDeleteMe (https://justdeleteme.xyz) to remove old accounts.

- Delete unused social media, email, and forum accounts.

Step 2: Opt-Out of Data Brokers

- Remove data from Spokeo, Whitepages, Acxiom, and PeopleFinder.

- Use Blur (Abine) or DeleteMe to automate opt-out requests.

Step 3: Remove Search Engine Indexing

- Request URL removals via Google's "Right to be Forgotten" (EU) or CCPA (US).

- Add a robots.txt file to prevent Google from crawling personal websites.

Step 4: Scrub Metadata from Files and Images

- Remove EXIF metadata from images before uploading:

```
1    exiftool -all= photo.jpg
```

- Strip PDF metadata:

```
1    pdfinfo -meta document.pdf
```

Step 5: Monitor for Data Leaks

- Set up alerts on HaveIBeenPwned to track leaked credentials.

- Use Google Alerts for your name, email, or username leaks.

Preventing Email Header & Metadata Leaks

Emails carry metadata including:

- Sender's IP address (especially with webmail or SMTP clients).

- Server timestamps and routing information.

- Email client fingerprints (e.g., Outlook, Thunderbird).

Mitigation:

- Use webmail providers like **ProtonMail** or **Tutanota** that strip sender IP metadata.

- Send emails over Tor or a VPN to mask originating IP.

- Review email headers with tools like mxtoolbox.com before sending sensitive content.

- Prefer end-to-end encrypted messaging for high-risk communications.

Leaking Through Metadata in Code Repositories

Open-source developers often unintentionally leak:

- **Git Commit Metadata:** Author names, email addresses, and timestamps.

- **Repository History:** Deleted files remain accessible through commit history.

- **Dotfiles and Credentials:** Accidentally committed '.env', '.bash_history', or API keys.

Mitigation:

- Use git config --global user.name and user.email with anonymous values.

- Scrub sensitive history with tools like BFG Repo-Cleaner.

- Use pre-commit hooks to block committing secrets.

27.3 Preventing Cross-Platform Stylometric Profiling

Adversaries can link accounts by writing style (stylometry):

- Sentence structure, grammar, and punctuation.

- Unique vocabulary and phrasing.

Mitigation:

- Vary writing styles or use AI paraphrasing tools.

- Avoid posting long-form content in multiple accounts.

- Review text with anonymization tools like **Anonymouth**.

27.4 Controlling Search Engine Caching and Archives

Old web pages may live on through caching and archiving services:

- Google Cache, Archive.org (Wayback Machine), and archive.today.

Mitigation:

- Request cache removal via Google's Removal Tool.

- Submit takedown requests to Archive.org.

- Prevent future archiving with the robots.txt directive:

```
1    User-agent: Archive.org_bot
2    Disallow: /
```

27.5 Using Honeytokens to Detect OSINT Abuse

Plant unique, traceable data (honeytokens) to detect data scraping or abuse:

- Unique email aliases (SimpleLogin, AnonAddy) per service.

- Fake user profiles with non-existent details.

- Canary URLs or files to monitor access attempts.

Malicious OSINT Honeytraps & Decoy Data Attacks

Not all OSINT is passively collected—attackers may plant baited content to lure you into self-identification.

Common Honeytraps:

- Fake job postings or "recruiter" profiles requesting resumes or personal details.

- Impersonation of journalists, activists, or community members to harvest information.

- Malware-laden "leaked" documents designed to target researchers.

Mitigation:

- Vet unsolicited requests carefully—check reputation via OSINT tools.

- Avoid downloading untrusted "leaks" without verification.

- Use sandboxed or air-gapped environments to open unknown files.

- Treat all inbound communication as suspect until proven otherwise.

27.6 Monitoring Leaks With Dark Web Search Tools

Surface and deep web search engines designed for threat hunting:

- **DarkSearch.io, Ahmia, OnionLand Search** for Tor sites.

- **Dehashed, IntelX** for credential leaks.

Mitigation:

- Regularly check for leaked usernames, passwords, and domains.

- Use breach notification services with custom alerting.

Preventing OSINT Correlation

Even if an account is anonymous, attackers can still correlate it to your real identity.

How to prevent identity linking:

- Use different usernames, emails, and devices for each identity.

- Never log into personal accounts from privacy-focused browsers (Tor, Mullvad).

- Avoid posting unique images that can be reverse-searched.

Using Burner Identities and Disposable Emails

- Create separate compartmentalized online identities for different activities.

- Use burner email addresses (SimpleLogin, ProtonMail, TutaNota) for registrations.

- Avoid linking social media accounts (Facebook, LinkedIn, GitHub, Reddit).

Spoofing and Misleading OSINT Tools

- Use Fawkes Cloaking Tool to prevent AI facial recognition.

- Randomize online behavior to avoid pattern tracking.

- Mask your real IP address using Tor or a multi-hop VPN.

27.7 Social Graph Poisoning to Break Profiling Engines

AI-driven OSINT tools often build "social graphs" mapping relationships, connections, and interactions across platforms.

Poisoning Techniques:

- Connect to random, unrelated profiles to generate false graph nodes.

- Interact with disinformation, parody, or nonsense accounts to degrade graph quality.

- Maintain duplicate or throwaway accounts that mimic real behavior but lead nowhere.

Objective:

- Overwhelm AI profiling engines with useless or misleading data.

- Make your real relationships indistinguishable from noise.

Third-Party OSINT Risks (Supply Chain & Employers)

Even if you maintain perfect personal OPSEC, attackers may exploit third-party connections.

Common Targets:

- Public staff directories on company websites.

- Press releases revealing personal or role-specific details.

- Vendor partnerships or supply chain announcements.

Mitigation:

- Encourage employers to limit public-facing personal information.

- Request anonymized public profiles where possible.

- Monitor third-party mentions using OSINT tools or Google Alerts.

27.8 Preventing Device Fingerprinting

Websites and services can track you by identifying your hardware and software fingerprint.

How They Track You:

- Screen resolution, browser version, and OS metadata.

- Installed fonts, plugins, and audio stack fingerprinting.

- WebGL and Canvas fingerprinting.

Mitigation:

- Use privacy browsers like **Tor Browser** or **Mullvad Browser** with fingerprint resistance.

- Disable JavaScript where practical.

- Use browser extensions like **CanvasBlocker** or **Trace**.

Preventing Hardware Fingerprinting and Device Tracking

Devices reveal more than you think. Hardware identifiers can betray your location or identity.

- **IMEI/MEID Laundering:** Avoid reusing the same mobile device or IMEI across different operations.

- **MAC Address Randomization:** Ensure Wi-Fi MAC randomization is enabled on all devices.

- **Laptop Serial Number Awareness:** Avoid registering devices with your real identity.

- **Peripheral Fingerprinting:** Be cautious of unique USB device IDs that can be logged by forensic tools.

- **Device Laundering:** Where legal, acquire second-hand hardware with no purchase records or use intermediary buyers.

27.9 Building an Anonymous Online Presence

If you need to remain anonymous while still maintaining an online presence, follow these steps:

1. Use a new, anonymous device (burner laptop, virtual machine, or Tails OS).

2. Only connect through privacy tools (Tor, Mullvad VPN, Whonix).

3. Never reveal real personal details, writing style, or location metadata.

4. Use disposable identities and compartmentalized online accounts.

5. Monitor OSINT tracking tools regularly to remove leaks.

Summary: OSINT tools can expose personal information from social media, data breaches, and search engines. To counteract this, use strong digital hygiene, metadata scrubbing, burner identities, and privacy-focused tools.

28.

Final OPSEC Measures

Even the most secure systems can be compromised if proper Operational Security (OPSEC) is not followed. Nation-state actors, corporate adversaries, and cybercriminals deploy advanced attack techniques that exploit human error, metadata leaks, and hardware vulnerabilities.

This section provides the most advanced defensive measures against surveillance, physical compromise, and OSINT threats.

28.1 Protecting Against Nation-State Attacks

Nation-state adversaries have access to:

- Zero-day exploits - Attacks on undiscovered software vulnerabilities.

- Firmware backdoors - Hardware implants and malicious BIOS modifications.

- AI-assisted surveillance - Machine learning

algorithms that track patterns.

- Supply chain compromises - Hardware intercepted before it reaches the user.

28.2 Advanced Attack Techniques

Attack Type	How It Works	Defense
TEMPEST	Captures electromagnetic emissions from a device	Use a Faraday cage, shielded cables
AirHopper	Uses GPU to transmit data via radio waves	Disable speakers, use white noise jammers
BadUSB	USB drives inject malware upon connection	Use only trusted USBs, disable autorun
Evil Maid Attack	Installs malware on a system when left unattended	Use tamper-evident seals, disk encryption
Supply Chain Attack	Malware embedded in hardware at the manufacturing stage	Buy devices from trusted sources, verify firmware integrity
Metadata Exploitation	Analyzes timestamps, locations, and connections in encrypted data	Use Tor, change metadata before sending files
AI-Assisted Surveillance	Machine learning analyzes communication patterns	Randomize behaviors, avoid repeated schedules

28.3 Physical Security for Devices

A compromised device allows attackers full access to encrypted data. Protect against physical attacks using the following methods:

- **Use a Faraday Bag or Box** – Prevents remote activation, tracking, and wireless exfiltration.

- **Enable Full-Disk Encryption (FDE)** – Ensures all data remains encrypted even if seized.

- **Apply Tamper-Evident Seals** – Detects unauthorized access to laptops and storage.

- **Use Anti-Forensic Techniques** – Store encrypted files inside hidden volumes (VeraCrypt hidden containers).

- **Air-Gap High-Security Systems** – Never connect mission-critical devices to external networks.

28.4 Mitigating Insider Threats and Betrayal Vectors

Even the best technical security can be compromised by someone you trust. Insider threats include malicious colleagues, coerced friends, or betrayed operational partners.

- **Need-to-Know Principle:** Never share more information than absolutely necessary, even with trusted contacts.

- **Compartmentalize Teams:** Isolate responsibilities and access between operational roles to reduce damage if one person is compromised.

- **Cross-Verification:** Require independent verification from multiple channels before trusting critical information.

- **Behavioral Monitoring:** Watch for signs of coercion, stress, or sudden shifts in behavior that may indicate compromise.

- **Rotate Contacts:** Periodically change contact methods and identities to avoid long-term dependency on a single trust point.

Implementing Plausible Deniability

Create dual-layer security where only benign data appears to be present.

Examples:

- **Hidden VeraCrypt volumes** inside visible ones.

- Decoy social media accounts with misleading information.

- False operational timelines or dummy devices.

Purpose:

- Provide **coercion-safe disclosures** that reveal nothing sensitive.

28.5 Securing the Hardware Supply Chain

Hardware can be compromised before you even open the box.

Supply Chain Threats:

- Nation-state interdiction and implants.

- Firmware tampering or pre-installed malware.

Secure Acquisition Strategies:

- Purchase from **privacy-respecting vendors** (e.g., Purism, Nitrokey).

- **Physically inspect** packaging for tamper evidence.

- Validate **firmware hashes** against vendor checksums.

28.6 Countering Behavioral Exploitation

Nation-state and corporate actors often manipulate behavior to bypass technical defenses.

Exploitation Techniques:

- Social Engineering – Manipulating targets to bypass security.

- Honeytraps – Creating emotional or relational leverage.

- Psychological Profiling – Targeting known behavioral weaknesses.

Defensive Strategies:

- Train to recognize social engineering attempts (pretexting, phishing, baiting).

- Avoid emotionally charged digital interactions with unknown actors.

- Regularly audit personal behavior for patterns exploitable by adversaries.

- Rotate online personas and physical routines to break targeting consistency.

Traffic Shaping & Decoy Traffic Generation

Even encrypted or Tor traffic can be flagged through traffic analysis.

Detection Risks:

• Traffic fingerprinting based on packet timing and size.

• Identifying Tor or VPN usage through volume and behavior analysis.

Countermeasures:

• Deploy traffic obfuscation tools like **obfs4** or **meek** with Tor.

• Use VPNs with packet size padding or randomized connection intervals.

• Generate decoy traffic to simulate normal browsing behavior.

• Randomize connection times and traffic volumes to blend into common patterns.

Reducing Digital Fingerprints

To maintain OPSEC, minimize your digital footprint by following these techniques:

1. Isolate Identities

• Use burner identities - Never reuse real-world personal information.

• Separate online personas - Keep different activities fully compartmentalized.

- Avoid linking accounts – Never reuse emails, phone numbers, or IP addresses across different identities.

2. Spoof Location and Traffic Analysis

- Use multi-hop VPNs (Mullvad, ProtonVPN) – Hides your real IP address.

- Route traffic through Tor or Tor Bridges – Prevents ISP monitoring.

- Randomize MAC addresses – Avoids network tracking.

3. Anonymize Communication

- Use privacy-focused messengers (Session, Briar, Matrix).

- Avoid phone-based apps – Prefer P2P decentralized chat.

- Use disposable SIM cards and email accounts for communication.

28.7 Secure Computing Environments

For individuals working with sensitive data in high-risk environments, the following setup provides maximum security:

1. Air-Gapped Operations

- Store encryption keys on an offline, air-gapped device.

- Never connect secure systems to the internet.

- Use data diodes or QR codes for one-way data transfer.

2. Hardened Secure OS

- Use Qubes OS for multi-layered compartmentalization.

- Boot from Tails OS (amnesic live system) for anonymous operations.

- Run applications inside Firejail or AppArmor sandboxes.

3. Secure Hardware and BIOS

- Use coreboot/Libreboot firmware for verified boot security.

- Disable Intel Management Engine (IME) and AMD PSP.

- Choose Libre-certified hardware (Purism, ThinkPad X230 with coreboot).

4. Self-Destructing Cryptography

- Store sensitive data inside VeraCrypt hidden volumes.

- Use Shamir's Secret Sharing Scheme (SSSS) for key fragmentation.

- Set up dead-man switches to erase encryption keys upon forced access.

5. Device Sanitization and Destruction

- Securely wipe storage before disposal:

```
1    sudo shred -n 5 -z /dev/sdX
```

- Physically destroy SSDs using:

  - Industrial degaussing machines for HDDs.
  - Drilling through NAND flash memory chips on SSDs.
  - Shredding USB drives and SD cards to prevent data recovery.

28.8 Anonymity Workflow for High-Security Operations

To maintain full anonymity, follow this multi-layered approach:

1. **Compartmentalize Activities**

   - Use separate laptops, VMs, or live OS instances for different tasks.
   - Never mix personal and operational accounts.

2. **Anonymous Internet Access**

   - Use Tails OS or Whonix with Tor and VPN chaining.
   - Never log into personal accounts or reused email addresses.

3. **Secure Messaging and Transactions**

   - Use Session or Matrix for encrypted communication.
   - Conduct cryptocurrency transactions using Monero (XMR) to avoid blockchain analysis.

4. Avoid Predictable Patterns

- Change online activity patterns frequently.
- Use different exit nodes and VPN servers for each session.

Why: Long-term operational security requires mental endurance. Burnout leads to mistakes.

Warning Signs:

- Fatigue from constant privacy routines.
- Cutting corners "just this once."
- Emotional or mental fatigue from high-stress environments.

Countermeasures:

- Schedule OPSEC "off days" where no sensitive operations are performed.
- Practice mindfulness or stress-reduction techniques.
- Rotate trusted team members to share operational load.
- Conduct regular self-assessments to recognize slipping discipline.

5. Regularly Monitor for OSINT Exposure

- Use HaveIBeenPwned and Google Alerts to detect personal data leaks.
- Conduct Google Dorking on yourself to check for exposed data.

Summary: OPSEC is a constant battle. To stay ahead of adversaries, deploy hardware security, anonymity workflows, and nation-state-level defenses.

29.

OPSEC Checklist

Personal Security Checklist

- **Use Tor with a VPN** (Double-hop privacy for anonymity).

- **Encrypt email and messaging** (ProtonMail, Tutanota, Session, Signal).

- **Enable Full-Disk Encryption (FDE)** on all devices.

- **Use privacy-hardened OS** (GrapheneOS, Qubes OS, Tails).

- **Never use personal details** for anonymous accounts.

- **Secure all passwords with a password manager** (Bitwarden, KeePassXC).

- **Enable Multi-Factor Authentication (MFA/2FA)** (preferably hardware keys like YubiKey).

High-Risk OPSEC for Activists, Journalists, and Whistleblowers

- Use burner phones & SIMs (Cash-purchased, no personal data linked).

- Never log into personal accounts from high-risk devices.

- Avoid centralized cloud services (Google, iCloud, Dropbox).

- Store sensitive files on air-gapped devices or Tails OS USB.

- Disable Bluetooth, Wi-Fi auto-connect, and fingerprint unlock.

- Assume all phone calls, SMS, and emails are intercepted.

- Use code words for secure, deniable communication.

- Use Faraday bags when carrying sensitive devices.

- Set up dead-man switches to wipe sensitive data if arrested.

Advanced Anti-Surveillance Measures

- **Detect IMSI Catchers** (Use SnoopSnitch, CellSpyCatcher).

- **Use MAC address randomization** (Prevents Wi-Fi tracking).

- **Evade AI-assisted tracking** (Randomize routines, alter gait).

- **Mask heat signatures** (Use mylar blankets against thermal drones).

- **Use digital noise generators** (Disrupt voice-activated bugs).

- **Encrypt USB drives** (VeraCrypt hidden volumes, Tails persistent storage).

- **Use offline transactions** (Bitcoin + Monero, cash purchases).

- **Employ misinformation tactics** (Fake digital footprint creation).

Forensic Countermeasures

- Remove EXIF metadata before uploading images (Use 'exiftool').

- Sanitize file timestamps (Modify with 'touch' command).

- Wipe storage devices before disposal:

```
sudo shred -n 5 -z /dev/sdX
```

- Destroy SSDs properly (Physical destruction of NAND chips).

- Use "duress passwords" (Triggers self-destruction of encryption keys).

30.

Additional Resources

For further learning and advanced security techniques, refer to the following:

30.1 Privacy-Focused Operating Systems

- https://www.digi77.com/linux-kodachi/ – For privacy and Security.

- https://www.grapheneos.org – Secure mobile OS (Android alternative).

- https://www.qubes-os.org – Compartmentalized security OS.

- https://www.torproject.org – Anonymous browsing over Tor.

- https://tails.net – Secure, amnesic live OS (ideal for journalists).

- https://www.whonix.org – Anonymous OS designed for Tor-based security.

Online Security & Anonymity Guides

- https://ssd.eff.org - Electronic Frontier Foundation's security guide.

- https://privacyguides.org - Up-to-date privacy tools & best practices.

- https://restoreprivacy.com - VPNs, private search engines, encrypted storage.

- https://prism-break.org - Open-source alternatives to mainstream tech.

30.2 Recommended Books on OPSEC and Digital Privacy

- **"Extreme Privacy: What It Takes to Disappear"** - Michael Bazzell.

- **"The Art of Invisibility"** - Kevin Mitnick.

- **"Practical OPSEC: Security and Privacy for Everyday Life"** - Justin Carroll, Drew M.

- **"Permanent Record"** - Edward Snowden (Whistleblower insights).

- **"Surveillance Countermeasures"** - ACM IV Security.

30.3 Cryptocurrency Privacy Tools

- https://www.getmonero.org - Monero (XMR) privacy-focused cryptocurrency.

- https://samouraiwallet.com - Bitcoin wallet with CoinJoin mixing.

- https://wasabiwallet.io - Bitcoin privacy wallet.

> **Final Note:** True privacy requires constant adaptation. OPSEC is an ongoing process, not a one-time setup. Always assume surveillance, and take proactive steps to minimize your exposure.

31.

Privacy Security Checklists

These checklists provide actionable steps for various threat models. Whether you are a journalist in a hostile country, a whistleblower, or a privacy-conscious individual, these guidelines will help protect your digital and physical security.

31.1 General Privacy User Checklist

- Use **Brave, Mullvad Browser, or Tor** instead of Chrome.

- Install **Mullvad VPN** or **iVPN**.

- Use **ProtonMail / Tutanota** instead of Gmail.

- Disable **all telemetry, tracking, and ad personalization**.

- Secure **social media** (fake info, privacy settings locked down).

- Use **Yubikey or physical security keys** for authentication.

- Encrypt **all hard drives and USBs** with **Veracrypt.**
- Stop using **Google Drive / Dropbox,** use **CryptPad, Proton Drive, or Nextcloud.**
- Use a **burner email** (SimpleLogin, AnonAddy) for online accounts.
- Use **Containerized Browsing** (e.g., Firefox Multi-Account Containers) to prevent cross-site tracking.
- Clear **browser cookies and site data** after every session.

31.2 Journalist / Activist in a Hostile Country

- Use an **offline laptop (Tails OS, Qubes OS)** for sensitive work.
- Carry a **Faraday bag** for your phone to prevent tracking.
- Never use **personal numbers;** get a **burner SIM.**
- Use **Briar, Signal, or Session** for communication.
- Set up a **dead man's switch** in case of disappearance.
- Avoid hotel Wi-Fi and never use the same network twice.
- Have a **USB kill switch** to erase data instantly if arrested.
- Store important files in **steganographic images/audio files.**

- Use **cash or Monero for purchases**, avoid any identity links.

- Keep **multiple copies of ID, cash, and emergency escape routes.**

- Carry **decoy devices** or **dual-boot systems**—harmless data on one partition, sensitive data on another.

- Use **dead-letter drops** (physical or digital) to pass information without real-time communication.

31.3 Whistleblower / Leaker Checklist

- Never leak from your home or personal internet.

- Use **Tor and Tails OS** for all leaks.

- Strip **metadata** from documents before sharing.

- Use **SecureDrop** to send info to journalists.

- Communicate **only through burner email accounts via Tor.**

- Store **all sensitive data on encrypted USBs**, not your laptop.

- Do not discuss leaks with **anyone**, even close contacts.

- Have a **plan for exposure**—fake accounts, an escape route, etc.

- Use **a fake identity** and keep **no personal records** of your actions.

- Validate file hashes before uploading or transferring documents to ensure they weren't tampered with.

31.4 Ex (Stalking/Abuse Prevention)

- Get a **new phone, new SIM, and new email.**

- Never reuse old passwords or email accounts.

- Lock down **social media**—change names, remove old posts, go private.

- Use **Life360 deception techniques** (dummy phone to mislead location tracking).

- Set up **Google Voice or MySudo** for phone calls instead of a real number.

- Get a **P.O. Box**—do not receive mail at your real address.

- Use **Cash App, Monero, or privacy-based payment methods**.

- Never use your home **Wi-Fi** for sensitive searches or messages.

- Install **surveillance detection apps** (Haven, Mic-Lock, etc.).

- Use **home security cameras, doorbell cams, and motion sensors**.

- **Erase all old data** from shared accounts.

- Regularly **scan your vehicle and home** for hidden GPS trackers or surveillance devices.

- Set up **privacy notifications** (e.g., alerts when new photos of you appear online using PimEyes).

31.5 High-Profile / Politician Privacy Checklist

- Have **dedicated security hardware** (secure phone, Faraday bag, encrypted storage).

- Only communicate via **offline devices** for sensitive topics.

- Use **multiple aliases & accounts**—never use real names on social media.

- Rotate **burner SIMs** weekly.

- Avoid **public Wi-Fi, smart home devices, and cloud services.**

- Always use **physical security keys** for account logins.

- Have **separate online personas** (one for business, one for personal).

- Use **AI voice-cloning detection tools** to prevent deepfake attacks.

- Secure home and office with **signal jammers & RF scanners.**

- Avoid **biometric authentication**—use **PINs, Yubikeys, or passcodes** only.

- Employ **media coaching** to control public narrative and minimize OSINT leaks during interviews.

- Use **encrypted calendar and scheduling apps** (e.g., Proton Calendar) to avoid timeline leaks.

31.6 Cryptocurrency & Financial Privacy Checklist

- Use **Monero (XMR) or Zcash** instead of Bitcoin.
- Mix coins with **CoinJoin (Wasabi, Samourai Wallet)**.
- Buy crypto with **cash or gift cards**, never a personal bank account.
- Use **non-KYC exchanges** like Bisq or local P2P trades.
- Store assets on **cold storage (offline)**.
- Never discuss crypto holdings on social media.
- Use **a dedicated, separate phone or laptop** for crypto transactions.
- Have a **"fake" crypto portfolio** in case of forced access.
- Rotate **multiple wallets** for separate transactions to prevent address clustering.
- Use **Tor or a VPN** when accessing crypto services to prevent IP address exposure.
- Split large transactions into **smaller, randomized amounts** to evade transaction fingerprinting.

These checklists serve as quick guides for various situations where privacy, security, and anonymity are crucial.